# Nitrogen Assessment

# Nitrogen Assessment
## Pakistan as a Case-Study

Edited by

**Tariq Aziz**

*University of Agriculture Faisalabad, Sub-Campus at Depalpur,
Okara, Punjab, Pakistan*

**Abdul Wakeel**

*Institute of Soil and Environmental Sciences,
University of Agriculture Faisalabad, Punjab, Pakistan*

**Muhammad Arif Watto**

*University of Agriculture Faisalabad, Sub-Campus at Depalpur,
Okara, Punjab, Pakistan*

**Muhammad Sanaullah**

*Institute of Soil and Environmental Sciences,
University of Agriculture Faisalabad, Punjab, Pakistan*

**Muhammad Aamer Maqsood**

*Institute of Soil and Environmental Sciences,
University of Agriculture Faisalabad, Punjab, Pakistan*

**Aysha Kiran**

*Department of Botany, University of Agriculture Faisalabad,
Punjab, Pakistan*

ELSEVIER

**ACADEMIC PRESS**

An imprint of Elsevier

Academic Press is an imprint of Elsevier
125 London Wall, London EC2Y 5AS, United Kingdom
525 B Street, Suite 1650, San Diego, CA 92101, United States
50 Hampshire Street, 5th Floor, Cambridge, MA 02139, United States
The Boulevard, Langford Lane, Kidlington, Oxford OX5 1GB, United Kingdom

**Library of Congress Cataloging-in-Publication Data**
A catalog record for this book is available from the Library of Congress

**British Library Cataloguing-in-Publication Data**
A catalogue record for this book is available from the British Library

ISBN: 978-0-12-824417-3

For information on all Academic Press publications visit our website at
https://www.elsevier.com/books-and-journals

*Publisher:* Megan Ball
*Acquisitions Editor:* Nancy Maragioglio
*Editorial Project Manager:* Devlin Person
*Production Project Manager:* Kiruthika Govindaraju
*Cover Designer:* Alan Studholme

Typeset by TNQ Technologies

# Contents

# Contributors

**Waqar Ahmad**
Asian Soil Partnership, and School of Agriculture and Food Sciences, The University of Queensland, Brisbane, QLD, Australia

**Imran Ashraf**
Institute of Soil and Environmental Sciences, University of Agriculture Faisalabad, Punjab, Pakistan

**Masood Iqbal Awan**
University of Agriculture Faisalabad, Sub-Campus at Depalpur, Okara, Punjab, Pakistan

**Tariq Aziz**
University of Agriculture Faisalabad, Sub-Campus at Depalpur, Okara, Punjab, Pakistan

**Zunaira Bano**
Department of Botany, University of Agriculture Faisalabad, Punjab, Pakistan

**Bijay-Singh**
Punjab Agricultural University, Ludhiana, Punjab, India

**Hafiz Muhammad Bilal**
Water Management Research Farm, Renala Khurd, Okara, Pakistan

**Amara Farooq**
Institute of Soil and Environmental Sciences, University of Agriculture Faisalabad, Punjab, Pakistan

**Ghulam Haider**
Department of Plant Biotechnology, Atta-ur-Rahman School of Applied Biosciences, NUST, Islamabad, Pakistan

**Nighat Hasnain**
Agrilenz Ltd., London, United Kingdom

**Muhammad Irfan**
Soil and Environmental Sciences Division, Nuclear Institute of Agriculture (NIA), Tandojam, Sindh, Pakistan

**Annie Irshad**
College of Grassland Agriculture, Northwest A&F University, Yangling, Shaanxi, China

**Aysha Kiran**
Department of Botany, University of Agriculture Faisalabad, Punjab, Pakistan

**Muhammad Aamer Maqsood**
Institute of Soil and Environmental Sciences, University of Agriculture Faisalabad, Punjab, Pakistan

**Abdul Jalil Marwat**
National Fertilizer Development Centre, Islamabad, Pakistan

**Fathia Mubeen**
Soil and Environmental Biotechnology Division, National Institute for Biotechnology and Genetic Engineering, Faisalabad, Pakistan

**Ahmad Mujtaba**
Institute of Soil and Environmental Sciences, University of Agriculture Faisalabad, Punjab, Pakistan

**Naqsh-e-Zuhra**
Institute of Soil and Environmental Sciences, University of Agriculture Faisalabad, Punjab, Pakistan

**Muhammad Nasim**
Pesticide Quality Control Laboratory, Bahawalpur, Punjab, Pakistan

**Allah Nawaz**
Institute of Soil and Environmental Sciences, University of Agriculture Faisalabad, Punjab, Pakistan; Soil Chemistry Section, Ayub Agricultural Research Institute, Faisalabad, Punjab, Pakistan

**Nasir Rasheed**
Institute of Soil and Environmental Sciences, University of Agriculture Faisalabad, Punjab, Pakistan

**Sajjad Raza**
School of Geographical Sciences, Nanjing University of Information Science & Technology, Nanjing, Jiangsu, China

**Robert Rees**
Scotland's Rural College (SRUC) Edinburgh, Edinburgh, United Kingdom

**Hafeez ur Rehman**
Department of Agronomy, University of Agriculture Faisalabad, Punjab, Pakistan

**Muhammad Sanaullah**
Institute of Soil and Environmental Sciences, University of Agriculture Faisalabad, Punjab, Pakistan

**Zia-ul-Hassan Shah**
Department of Soil Science, Sindh Agriculture University, Tandojam, Sindh, Pakistan

**Muhammad Rizwan Shahid**
Institute of Soil and Environmental Sciences, University of Agriculture Faisalabad, Punjab, Pakistan

**Ahmad Naeem Shahzad**
Department of Agronomy, Bahauddin Zakariya University, Multan, Pakistan

**Mark Sutton**
Centre for Ecology and Hydrology, Edinburgh Research Station, Penicuik, Midlothian, United Kingdom

**Abdul Wakeel**
Institute of Soil and Environmental Sciences, University of Agriculture Faisalabad, Punjab, Pakistan

**Muhammad Arif Watto**
University of Agriculture Faisalabad, Sub-Campus at Depalpur, Okara, Punjab, Pakistan

**Xiaoning Zhao**
School of Geographical Sciences, Nanjing University of Information Science & Technology, Nanjing, Jiangsu, China

**Munir Hussain Zia**
R&D Department, Fauji Fertilizer Company Limited, Rawalpindi, Punjab, Pakistan

# Foreword

For the past 50 years or so, global society has taken nitrogen too much for granted. A look at chemistry textbooks from the 1920s will point to a past "nitrogen problem," where a growing human population would require more food. To boost crop yields and produce this food would need increased access to limited nitrogen compounds and other nutrients. Recycling of organic nitrogen sources in manure and biological nitrogen fixation was apparently not enough, while mining of "fossil nitrogen" either as guano or saltpeter risked quickly depleting limited stocks. With the advent of the Haber–Bosch process, it seemed that the nitrogen problem had been solved, since this provides a cheap way to harvest atmospheric di-nitrogen ($N_2$) and turn it into ammonia ($NH_3$) for fertilizer and other uses. The net result was a massive increase in chemical nitrogen fertilizers, especially from the 1950s, providing the fuel for the Green Revolution, where new high-demand, high-yield crop varieties were the engine.

As we see now, this was only the start of a new kind of nitrogen problem. It turned out that nitrogen use is rather inefficient, meaning that a large amount of Haber–Bosch nitrogen compounds is lost to the environment. At the same time, it was realized that nitrogen is not just an agricultural challenge. Burning fossil fuels and biofuels releases stored nitrogen as nitrogen oxides ($NO_x$), nitrous oxide ($N_2O$), and ammonia to the atmosphere, while high temperature combustion also converts atmospheric di-nitrogen into $NO_x$ and these other nitrogen pollutants. Apart from agriculture, industry, energy, and transport, the whole food system is implicated, as wastewater systems release nitrogen to groundwaters, water courses, and the coastal zone, threatening drinking water, ecosystems, and fisheries. Nitrogen compounds emitted by human activities to the atmosphere contribute to air pollution, climate change, and stratospheric ozone depletion, affecting our health, livelihoods, and ecosystems. And it is not just aquatic ecosystems that are affected. When nitrogen air pollution lands on forest, mountains, and other natural habitats, the ecology is changed, affecting ecosystem resilience and compromising ecosystem services.

In this brief summary it quickly becomes clear why we say that nitrogen is "everywhere and invisible." We face a problem of science/policy fragmentation across issues, where nitrogen is relevant for all of the 17 United Nations Sustainable Development Goals (SDGs) and yet is almost completely missing from the SDG process. To accelerate progress, we need to bring nitrogen together. We need to show how nitrogen links all aspects of our lives, all aspects of global change, and that nitrogen action needs to be embedded throughout the sustainable development agenda.

With this thinking in mind, I am delighted to welcome publication of the *Pakistan Nitrogen Assessment*, which brings together multiple strands of evidence of why nitrogen is relevant, and the opportunities for action. Under the guidance of leading researchers from the Pakistan scientific community, I expect that it will play a key role in raising awareness of how nitrogen management is essential if we are to reach the SDGs.

We have referred to the run up to 2030 as the "Nitrogen Decade," as highlighted during World Environment Day 2021, hosted by the Government of Pakistan in launching the "UN Decade of Ecosystem Restoration." The importance for sustainable ecosystems is obvious. At the same time actions for sustainable nitrogen management—following up the UN Environment Assembly Resolution 4/14 and the Colombo Declaration—will help to meet multiple economic, health, and environmental goals, supporting food and energy production, climate resilience, livelihoods, and the circular economy.

The solutions and synergies described in this book demonstrate Pakistan's leadership, and will provide welcome guidance for other countries as the world starts to embrace the new nitrogen challenge.

**Mark Sutton**
*Director, GEF/UNEP International Nitrogen Management System (INMS)*
*Director, GCRF South Asian Nitrogen Hub*
*Co-chair, UNECE Task Force on Reactive Nitrogen*

# Acknowledgments

The Editorial Team would like to extend sincere gratitude to all contributors especially to the lead authors for their valuable contribution and support. It's been a long journey since the inception of this great idea in 2019 during the South Asia Nitrogen Hub (SANH) meeting at Nepal, in February 2019, to the receipt of final draft. Thanks again!!

We are also highly indebted to University of Agriculture, Faisalabad, being the parent institute and partner of SANH from Pakistan.

We also owe very special thanks to SANH established under the UKRI-GCRF.

**This contribution is dedicated to farmers "the unsung heroes of war against hunger."**

**Tariq Aziz, Abdul Wakeel, M. Arif Watto, M. Aamer Maqsood, Muhammad Sanaullah and Aysha Kiran**

# Rethinking nitrogen use: need to plan beyond present

1

**Tariq Aziz[1], Abdul Wakeel[2], Ahmad Naeem Shahzad[3], Robert Rees[4], Mark Sutton[5]**

[1]*University of Agriculture Faisalabad, Sub-Campus at Depalpur, Okara, Punjab, Pakistan;*
[2]*Institute of Soil and Environmental Sciences, University of Agriculture Faisalabad, Punjab, Pakistan;* [3]*Department of Agronomy, Bahauddin Zakariya University, Multan, Pakistan;*
[4]*Scotland's Rural College (SRUC) Edinburgh, Edinburgh, United Kingdom;* [5]*Centre for Ecology and Hydrology, Edinburgh Research Station, Penicuik, Midlothian, United Kingdom*

## 1.1 Reactive nitrogen: a global challenge

In the early 1900s, food production was constrained by depleting soil nitrogen (N) stocks. This N constraint was unlocked by Fritz Haber and Carl Bosch by converting the inert dinitrogen gas into readily available N and as per recent estimates global food production may be reduced to one-half of the current level without N fertilizer applications (Erisman et al., 2008). The need for N use is a double-edged sword. On the one hand, its use is crucial in reducing poverty and hunger and boosting economic development, it can impose risks to health, environment, and economy on the other hand. The risks include climate change (Davidson, 2009), air pollution causing cancer and respiratory diseases in humans (Townsend et al., 2003), water pollution causing eutrophication in water bodies (Diaz and Rosenberg, 2008), loss of biodiversity (Clark and Tilman, 2008), stratospheric ozone depletion (Ravishankara et al., 2009), and acidification of natural ecosystems (Driscoll et al., 2003). The N compounds such as $NH_3$ and nitrous oxides form atmospheric aerosols that can cause significant economic and health damages (Paulot and Jacob, 2014).

Nitrogen pollution is one of the major environmental issues of the 21st century. Several sources (agriculture, biomass burning, energy and transport, and wastewater treatment) contribute to N pollution, of which agriculture is the major contributor with about 66% share in the total N emissions. Due to inefficiency and poor management, more than one-half of the inorganic fertilizers and manures added to soil eventually end up polluting the environment in one or the other way (Sutton et al., 2013). In addition to synthetic fertilizers, livestock supply chains also impact N emissions and global N flows (Uwizeye et al., 2020). It is estimated that the livestock sector emits one-third of the global anthropogenic N emissions equivalent to $65 \text{ Tg N yr}^{-1}$ (Uwizeye et al., 2020).

Nitrogen Assessment. https://doi.org/10.1016/B978-0-12-824417-3.00007-1

The current human interventions in the N cycle are contributing to increasing environmental impacts and could lead to irreversible change as these have already crossed the planetary boundaries (Steffen et al., 2015). One of the key N polluting compounds is nitrous oxide ($N_2O$), which contributes to 6% of the global annual emissions of greenhouse gases (Blanco et al., 2014). Nitrous oxide has 265 times more potential to cause global warming than carbon dioxide (Tian et al., 2019). Mitigating N pollution will therefore benefit the environment not only in terms of direct climate benefits to limit the increases in global temperatures but also in indirect climate benefits primarily from avoided pollution to the water bodies and atmosphere (Brink and van Grinsven, 2011; Kanter et al., 2017).

## 1.2 Nitrogen cycle and human intervention

The Earth's N largely exists in the form of dinitrogen ($N_2$; 99.9%) and is not directly available to most organisms. To convert ("fix") N into useable forms, high temperature and pressures (combustion or Haber–Bosch) or microbial activity (biological N fixation) are required to break the N triple bond. The fixed or reactive N (Nr) that is formed can then be transformed into several forms of inorganic [ammonium ($NH_4^+$), ammonia ($NH_3$), nitrate ($NO_3^-$), nitrous oxide ($N_2O$), nitric oxides ($NO_x$), and nitric acid ($HNO_3$)] and organic compounds (urea, proteins, amines, and nucleic acids). Human activities such as agriculture and the burning of fossil fuels release Nr into the environment, where it can be transformed into different chemical forms and stay in the environment for extended periods. The Nr keeps circulating in the environment with multiple effects on the atmosphere, fresh and marine waters, and on soils in a sequence of effects described as the "nitrogen cascade" until it is converted back into $N_2$ (Galloway et al., 2003). When a unit of synthetic fertilizer N is applied to croplands, it can be converted into various chemical forms and then either taken up by plants or lost to the environment. For example, from a unit of applied fertilizer, 30% could be lost by leaching as $NO_3^-$ causing eutrophication in water bodies, 10% could be lost in the form of $NO_x$ and $NH_3$ damaging air quality, and 1.3% could be directly or indirectly lost as $N_2O$ contributing to climate change and ozone depletion (Kanter, 2018). A 100 kg of N applied to 1-ha land in the United States could cost about $1716 for environmental damage to society (Kanter et al., 2015). The biggest damage cost comes from $NO_3^-$ (86%) pollution of water and $NO_x$ (8%) and $NH_3$ (5%) pollution of air, while damage in terms of climate change and ozone depletion from $N_2O$ is less than 1% (Kanter, 2018). This suggests that environmental damages from nonclimate components of N pollution are much higher and require more policy focus. Nevertheless, mitigating nonclimate N pollution will ultimately help mitigate the $N_2O$-related climate changes as well. Reducing global N pollution by encouraging nonclimate actions that could deliver climate benefits is one of the key strategies under the building block framework of the Paris Climate Agreement (Fig. 1.1) and is a way toward reducing 50% of N waste by 2030 as agreed by Columbo Declaration in 2019.

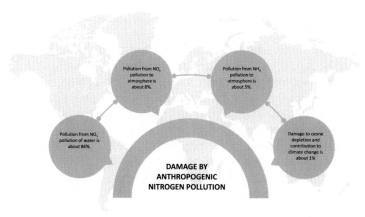

**FIGURE 1.1**

Contribution to environmental pollution through Nr boosted by human activity (Kanter, 2018).

A broad and integrated approach is needed in making policies to mitigate climate and nonclimate impacts of N pollution as mitigation of one component of N pollution (for example, efforts to reducing $NO_3$ leaching and runoff) can exacerbate emission of other components of the N cascade ($NH_3$, $NO_x$, and $N_2O$ emissions) (Kanter, 2018). However, the environmental damage caused by N losses needs to be balanced against the benefits that are associated with N use in agriculture and the role it plays in contributing to food security and economic development. It is therefore important to account for the costs and benefits of mitigation policies at local, regional, and global levels.

## 1.3 Nitrogen use efficiency in agriculture

Nitrogen use efficiency (NUE) is a critical indicator for N use in developing sustainable food security targets with less degradation of the environment (Galloway et al., 2008; Zhang et al., 2015). NUE is a measure of the amount of N recovered in the harvest expressed as a proportion of the amount of N supplied as synthetic fertilizer, manure, biological fixation, and atmospheric deposition. Nitrogen inputs and the corresponding NUEs vary significantly between countries and regions (Conant et al., 2013; Shahzad et al., 2019). Several studies have estimated an overall global NUE 42%−45% with a total global N input of 163−174 Tg N $yr^{-1}$ in 2009−10 (Zhang et al., 2015; Lasaletta et al., 2014; Mueller et al., 2017; Shahzad et al., 2019). During the past three decades, several developed nations (Europe, Canada, and the United States) have demonstrated a significant improvement in NUE, while many developing countries (Asia particularly India, China, and Pakistan) persistently recorded a little or no change in NUE (Conant et al., 2013). A recent analysis of N use in 124 countries found that NUE in Pakistan has drastically decreased from 53% in 1961 to 21% in 2009 reflecting the increased use of N fertilizers (Lassaletta et al., 2014). However, in regions such as Europe and North America new management approaches have helped increase NUE in the recent decades despite relatively high N fertilizer use (Omara et al., 2019). The current levels of NUE across countries and regions are impacted by several factors such as N input rates, balanced fertilization, soil, and climatic conditions, cropping systems, adoption of new technology and best management practices, socioeconomic development status, and introduction of environmental regulations (Zhang et al., 2015; Davidson et al., 2015). Many countries of Western Europe have achieved significant gains in NUE by introducing regulations aimed at limiting N input rates particularly in circumstances where significant losses are anticipated (van Grinven et al., 2012).

The high-input and low-output cropping systems in South Asia (Pakistan and India) undermine any effort to improve NUEs (Shahzad et al., 2019, Fig. 1.2). In recent years, increases in fertilizer prices as compared to crop prices reduced the application of N inputs and improved NUE in France and the United States, while increased subsidies on fertilizers in India, China, and Pakistan have yielded opposite results (Zhang et al., 2015; Shahzad et al., 2019). Imbalanced fertilization is also a big impediment to achieving the food security targets at sustainable NUEs. For example, little or no application of potassium (K) fertilizers to exhaustive crops could be responsible for extremely low NUEs in Pakistan (Shahzad et al., 2019). NUE is also reduced when fertilizer applications fail to take account of preexisting soil N supplies leading to significant overfertilization and potential N loss (Song et al., 2018). Depletion of soil carbon and phosphorus could also reduce NUE (Reis et al., 2016; Zingore et al., 2007). Improvement in the socioeconomic status also creates environmental awareness that could result in sustainable intensification with improved NUE (Zhang et al., 2015). In contrast to developed countries, large parts

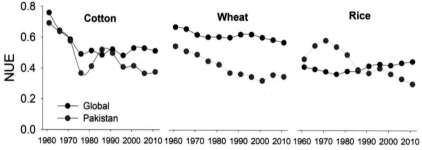

**FIGURE 1.2**

Comparison of NUE between global versus Pakistan during 1961–2014 for cotton, wheat, and rice.

*Modified from Shahzad, A.N., Qureshi, M.K., Wakeel, A., Misselbrook, T., 2019. Crop production in Pakistan and low nitrogen use efficiencies. Nat. Sustain. 2, 1106–1114.*

of Africa and some parts of Asia and Latin America do not have access to affordable N fertilizers (Austin et al., 2013; Vitousek et al., 2009), where crops grown on low inputs are depleting the soil nutrient and carbon stocks (Zingore et al., 2007). Since most of the management measures of N inputs and manures occur at the farm level, creating awareness and implementing any regulation at the farm level seems nearly impossible. In this context, regulating the fertilizer industry to gradually introduce more efficient fertilizer products such as slow-release fertilizers can offer significant opportunities for improving NUE.

## 1.4 Pakistan N budget: past, present, and future perspectives

Pakistan has a rapidly growing population of 220 million that is highly vulnerable to climate change and to a further worsening food security situation. The National Nutrition Survey in 2018 conducted by the United Nation's World Food Program shows that currently 36.9% of the population faces food insecurity (UNWFP, 2020). Given the food security situation, agricultural production is needed to be enhanced for meeting future food demands but in a sustainable and environment-friendly way.

The N budget of Pakistan has recently been calculated by Raza et al. (2018) who reported a significant increase in N fertilizer use from 1961 to 2013. Over this period, the NUE has decreased significantly, and the current NUE estimates report that NUE seldom exceeds 45% (Raza et al., 2018; Shahzad et al., 2019). In 2014−15, about 3.2% of the world's total chemical N resources were used on the irrigated landscape of Pakistan (Heffer et al., 2017). The increased N fertilizer use in Pakistan can also be linked to fertilizer subsidies by the Government. In recent years, there has been a reduction in general sales tax on urea from 17% to 5% and Rs. 100−200 subsidy on a 50 kg bag of urea apart from the voluntary price reduction by the fertilizer manufacturers that has led to an increase in cumulative sales and consequently the consumption of urea fertilizer from 40 to 63 million tons in 2018 (NFDC 2019).

The livestock sector in the country also plays a significant role in N input use and losses (Fig. 1.3). However, the NUE of the livestock sector is seriously low. Nitrogen is lost to the environment because of poor manure management (leaching and volatilization) and enteric fermentation. It has been estimated that optimizing the use of livestock and other organic wastes within Pakistan could reduce synthetic fertilizer N requirements by 43% without any loss of yield and providing financial benefits to farmers (Akram et al., 2018).

Pakistan is the world's sixth most populous country with significant growth in the economy; still, its share in the world's N emissions is negligible. In Pakistan, most of the N emissions are from the agriculture sector, reflecting low NUE by both crop production and livestock farming including manure management (Akram et al.,

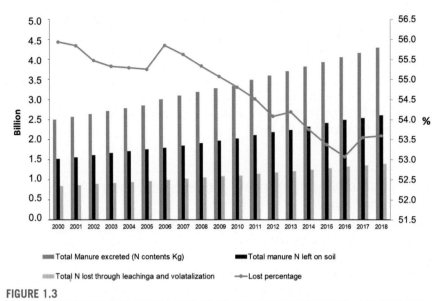

**FIGURE 1.3**

Role of livestock in N-related budget in Pakistan (FAO Dataset, 2020).

2019). Emissions from the transport sector, the burning of fossil fuels, energy generation, and industries have also been reported as contributors to N pollution, but their share is very small when compared with the agriculture sector which collectively produced 78.12 Tg $CO_{2eq}$ of $N_2O$ during 2016 (Mir et al., 2017). In another study (Ijaz and Goheer, 2021), the total emission from the agriculture sector is estimated at 83 Mt $N_2O- CO_{2eq}$.

Therefore, the current scenario of N fertilizer use in Pakistan needs serious efforts to revamp the agricultural interventions to make it more sustainable and environment friendly. Scientific and technological advancements offer various options to decrease N losses; however, government policies always play a critical role in the adoption of scientific interventions and adaptations to technological use worldwide.

## 1.5 Global attention toward nitrogen challenge

The N cycle involves a complex range of biogeochemical processes which regulate the reactive N at different spatial and temporal scales. Nitrogen from the atmosphere is fixed into reactive N through electric discharges, industrial and biological processes. The reactive N then undergoes multiple transformations: from inorganic to organic and vice versa in soil, assimilated by plants and microbes, and moves to animals and humans, before being released to air and water bodies and back to the atmosphere. Human interventions (increased fertilization, industrialization, fossil fuel burning, urbanization, changes in land use, and transport of food and feed) have greatly influenced the N cycle as the total amount of reactive N has been doubled in the biosphere since the last century and is projected to increase further in the coming decades (Fowler et al., 2015). Only a small fraction of the reactive N is retained by plants and animals and the excess is released to soil, water, and air causing potentially serious environmental consequences including air and water pollution, stratospheric ozone depletion, loss of biodiversity, and ecosystem restoration (Sutton et al., 2017). The N pollution is underestimated in many parts of the world including South Asia, where N problems and possible solutions have so far received little attention. However, N pollution has gained the world's attention during 2019, when the Colombo Declaration to halve N waste as a part of the International Action Plans by 2030 was made. Similarly, on the December 18, 2020, the UNECE Convention of Long-range Transboundary Air Pollution has adopted a Guidance Document on Integrated Sustainable Nitrogen Management. This document adds to the guidance document on preventing and abating ammonia emissions from agricultural sources, which was adopted in 2012 (UNECE 2021). Policies aiming at actual assessment of the N budgeting and increased NUE are direly needed to combat the N pollution and to make the ambition of reducing the N waste to half by 2030.

## 1.6 Concluding remarks

Nitrogen is an important plant nutrient and a constituent of many biomolecules that are critical for food security. The N use has tremendously been increased during the last five decades around the globe. The low NUE results in a huge amount of surplus N, which ultimately escapes to the environment damaging soil, air, and water quality, causing global warming, and having deleterious effects on biodiversity and the ecosystem. Nitrogen assessment is direly needed to develop better management strategies on a global basis. The N pollution is one of the biggest challenges of the 21st century and recently UN Environment has adopted Columbo Declaration (2019) with the ambition to reduce N waste to 50%. In South Asia (India, Pakistan, and Bangladesh), the agriculture sector is the major contributor toward increased N emissions along with the burning of fossil fuels and biomass. Policies aimed at increased NUE, particularly in agriculture, are need of the time. The major drivers, sources and sinks of N, trends in N use, N emissions, and various strategies to cope with the N challenge have been discussed in later chapters, with special emphasis on Pakistan.

## References

Akram, U., Metson, G.v.S., Quttineh, N.H., Wennergren, U., 2018. Closing Pakistan's yield gaps through nutrient recycling. Front. Sustain. Food Syst. 2, 24.

Akram, U., Quttineh, N.H., Wennergren, U., Tonderski, K., Metson, G.v.S., 2019. Enhancing nutrient recycling from excreta to meet crop nutrient needs in Sweden GÇô a spatial analysis. Sci. Rep. 9, 10264.

Austin, A.T., Bustamante, M.M.C., Nardoto, G.B., Mitre, S.K., Pérez, T., Ometto, J.P.H.B., Ascarrunz, N.L., Forti, M.C., Longo, K., Gavito, M.E., Enrich-Prast, A., Martinelli, L.A., 2013. Latin America's nitrogen challenge. Science 340 (6129), 149.

Blanco, G., Gerlagh, R., Suh, S., Barrett, J., de Coninck, H.C., Diaz Morejon, C.F., Mathur, R., Nakicenovic, N., Ofosu Ahenkora, A., Pan, J., Pathak, H., Rice, J., Richels, R., Smith, S.J., Stern, D.I., Toth, F.L., Zhou, P., 2014. Drivers, trends and mitigation. In: Edenhofer, O., Pichs-Madruga, R., Sokona, Y., Farahani, E., Kadner, S., Seyboth, K., Adler, A., Baum, I., Brunner, S., Eickemeier, P., Kriemann, B., Savolainen, J., Schlömer, S., von Stechow, C., Zwickel, T., Minx, J.C. (Eds.), Climate Change 2014: Mitigation of Climate Change. Contribution of Working Group III to the Fifth Assessment Report of the Intergovernmental Panel on Climate Change. Cambridge University Press, Cambridge.

Brink, C., van Grinsven, H., 2011. Costs and benefits of nitrogen in the environment. In: Sutton, M.A., Howard, C.M., Erisman, J.W., Billen, G., Bleeker, A., Grennfelt, P., van Grinsven, H., Grizzetti, B. (Eds.), European Nitrogen Assessment. Cambridge Univ. Press, Cambridge.

Clark, C.M., Tilman, D., 2008. Loss of plant species after chronic low-level nitrogen deposition to prairie grasslands. Nature 451 (7179), 712–715.

Conant, R.T., Berdanier, A.B., Grace, P.R., 2013. Patterns and trends in nitrogen use and nitrogen recovery efficiency in world agriculture. Global Biogeochem. Cycles 27, 558–566.

Davidson, E.A., 2009. The contribution of manure and fertilizer nitrogen to atmospheric nitrous oxide since 1860. Nat. Geosci. 2, 659.

Davidson, E.A., Suddick, E.C., Rice, C.W., Prokopy, L.S., 2015. More food, low pollution (Mo Fo Lo Po): a grand challenge for the 21st century. J. Environ. Qual. 44, 305–311.

Diaz, R.J., Rosenberg, R., 2008. Spreading dead zones and consequences for marine ecosystems. Science 321 (5891), 926–929.

Driscoll, C.T., Whitall, D., Aber, J., Boyer, E., Castro, M., Cronan, C., Christine, L.G., Groffman, P., Hopkinson, C., Lambert, K., Lawrence, G., Ollinger, S., 2003. Nitrogen pollution in the northeastern United States: sources, effects, and management options. Bioscience 53 (4), 357–374.

Erisman, J.W., Sutton, M.A., Galloway, J.N., Klimont, Z., Winiwarter, W., 2008. How a century of ammonia synthesis changed the world. Nat. Geosci. 1, 636–639.

FAO Dataset, 2020. Livestock Manure. Food and Agriculture Organization of the United Nations. http://www.fao.org/faostat/en/#data/EMN.

Fowler, D., Steadman, C.E., Stevenson, D., Coyle, M., Rees, R.M., Skiba, U.M., Sutton, M.A., Cape, N., Dore, A.J., Vieno, M., 2015. Effects of global change during the 21st century on the nitrogen cycle. Atmos. Chem. Phys. Discuss. 15, 1747–1868.

Galloway, J.N., Aber, J.D., Erisman, J.W., Seitzinger, S.P., Howarth, R.W., Cowling, E.B., Cosby, B.J., 2003. The nitrogen cascade. Bioscience 53, 341–356.

Galloway, J.N., Townsend, A.R., Erisman, J.W., Bekunda, M., Cai, Z., Freney, J.R., Martinelli, L.A., Seitzinger, S.P., Sutton, M.A., 2008. Transformation of the nitrogen cycle: recent trends, questions, and potential solutions. Science 320, 889–892.

Kanter, D.R., Zhang, X., Mauzerall, D.L., 2015. Reducing nitrogen pollution while decreasing farmers' costs and increasing fertilizer industry profits. J. Environ. Qual. 44, 325–335.

Kanter, D.R., Wentz, J.A., Galloway, J.N., Moomaw, W.R., Winiwarter, W., 2017. Managing a forgotten greenhouse gas under existing U.S. law: an interdisciplinary analysis. Environ. Sci. Pol. 67, 44–51.

Heffer, P., Gruère, A., Roberts, T., 2017. Assessment of fertilizer use by crop at the global level 2014-2014/15. International Fertilizer Industry Association and International Plant Nutrition Institute.

Ijaz, M., Goheer, M.A., 2021. Emission profile of Pakistan's agriculture: past trends and future projections. Environ. Dev. Sustain. 23 (2), 1668–1687.

Kanter, D.R., 2018. Nitrogen pollution: a key building block for addressing climate change. Clim. Change 147, 11–21.

Lassaletta, L., Billen, G., Grizzetti, B., Anglade, J., Garnier, J., 2014. 50-year trends in nitrogen use efficiency of world cropping systems: the relationship between yield and nitrogen input to cropland. Environ. Res. Lett. 9, 105011.

Mir, K.A., Purohit, P., Mehmood, S., 2017. Sectoral assessment of greenhouse gas emissions in Pakistan. Environ. Sci. Pollut. Res. 24 (35), 27345–27355.

Mueller, N.D., Lassaletta, L., Runck, B.C., Billen, G., Garnier, J., Gerber, J.S., 2017. Declining spatial efficiency of global cropland nitrogen allocation. Global Biogeochem. Cycles 31, 245–257.

NFDC, 2019. Statistics national fertilizer development centre. http://www.nfdc.gov.pk/stat. html. (Accessed 15 May 2019).

Omara, P., Aula, L., Oyebiyi, F., Raun, W.R., 2019. World cereal nitrogen use efficiency trends: review and current knowledge. Agrosyst. Geosci. Environ. 2, 180045.

Paulot, F., Jacob, D.J., 2014. Hidden Cost of U.S. Agricultural exports: particulate matter from ammonia emissions. Environ. Sci. Technol. 48 (2), 903–908.

Ravishankara, A.R., Daniel, J.S., Portmann, R.W., 2009. Nitrous oxide ($N_2O$): the dominant ozone-depleting substance emitted in the 21st century. Science 326 (5949), 123–125.

Raza, S., Zhou, J., Aziz, T., Afzal, M.R., Ahmed, M., Javaid, S., Chen, Z., 2018. Piling up reactive nitrogen and declining nitrogen use efficiency in Pakistan: a challenge not challenged (1961–2013). Environ. Res. Lett. 13 (3), 034012.

Reis, S., Bekunda, M., Howard, C.M., Karanja, N., Winiwarter, W., Yan, X., Bleeker, A., Sutton, M.A., 2016. Synthesis and review: tackling the nitrogen management challenge: from global to local scales. Environ. Res. Lett. 11 (12), 120205.

Shahzad, A.N., Qureshi, M.K., Wakeel, A., Misselbrook, T., 2019. Crop production in Pakistan and low nitrogen use efficiencies. Nat. Sustain. 2, 1106–1114.

Song, X., Liu, M., Ju, X., Gao, B., Su, F., Chen, X., Rees, R.M., 2018. Nitrous oxide emissions increase exponentially when optimum nitrogen fertilizer rates are exceeded in the North China plain. Environ. Sci. Technol. 52, 12504–12513.

Steffen, W., Richardson, K., Rockstrom, J., Cornell, S.E., Fetzer, I., Bennett, E.M., Biggs, R., Carpenter, S.R., de Vries, W., de Wit, C.A., Folke, C., Gerten, D., Heinke, J., Mace, G.M., Persson, L.M., Ramanathan, V., Reyers, B., Sörlin, S., 2015. Planetary boundaries: guiding human development on a changing planet. Science 347, 736.

Sutton, M.A., Howard, C.M., Bekunda, M., Grizzetti, B., de Vries, W., van Grinsven, H.J.M., Abrol, Y.P., Adhya, T.K., Billen, G., Davidson, E.A., Datta, A., Diaz, R., Erisman, J.W., Liu, X.J., Oenema, O., Palm, C., Raghuram, N., Reis, S., Scholz, R.W., Sims, T., Westhoek, H., Zhang, F.S., 2013. Our Nutrient World: The Challenge to Produce More Food and Energy with Less Pollution. Centre for Ecology and Hydrology, Edinburgh, UK (on behalf of the Global Partnership on Nutrient Management and the International Nitrogen Initiative).

Sutton, M.A., Drewer, J., Moring, A., Adhya, T.K., Ahmed, A., Bhatia, A., Brownlie, W., Dragosits, U., Ghude, S.D., Hillier, J., Hooda, S., Howard, C.M., Jain, N., Kumar, D., Kumar, R.M., Nayak, D.R., Neeraja, C.N., Prasanna, R., Price, A., Ramakrishnan, B.,

Reay, D.S., Singh, R., Skiba, U., Smith, J.U., Sohi, S., Subrahmanyan, D., Surekha, K., van Grinsven, H.J.M., Vieno, M., Voleti, S.R., Pathak, H., Raghuram, N., 2017. The Indian nitrogen challenge in a global perspective. In: The Indian Nitrogen Assessment. Elsevier, pp. 9–28.

Tian, H., Yang, J., Xu, R., Lu, C., Canadell, J.G., Davidson, E.A., Jackson, R.B., Arneth, A., Chang, J., Ciais, P., Gerber, S., Ito, A., Joos, F., Lienert, S., Messina, P., Olin, S., Pan, S., Peng, C., Saikawa, E., Thompson, R.L., Vuichard, N., Winiwarter, W., Zaehle, S., Zhang, B., 2019. Global soil nitrous oxide emissions since the preindustrial era estimated by an ensemble of terrestrial biosphere models: magnitude, attribution, and uncertainty. Global Change Biol. 25 (2), 640–659.

Townsend, A.R., Howarth, R.W., Bazzaz, F.A., Booth, M.S., Cleveland, C.C., Collinge, S.K., Dobson, A.p., Epstein, P.R., Holland, E.A., Keeney, D.R., Mallin, M.A., Rogers, C.A., Wayne, P., Wolfe, A.H., 2003. Human health effects of a changing global nitrogen cycle. Front. Ecol. Environ. 1 (5), 240–246.

UNECE 2021. Guidance document on integrated sustainable nitrogen management. Agriculture, Food and Environment. https://unece.org/sites/default/files/2021-08/ECE_EB.AIR_149-2104922E_0.pdf.

UNWFP, 2020. World Food Program: Pakistan country brief. UNWF Country Brief. https://www.wfp.org/countries/pakistan.

Uwizeye, A., de Boer, I.J.M., Opio, C.I., Schulte, R.P.O., Falcucci, A., Tempio, G., Teillard, F., Casu, F., Rulli, M., Galloway, J.N., Leip, A., Erisman, J.W., Robinson, T.P., Steinfeld, H., Gerber, P.J., 2020. Nitrogen emissions along global livestock supply chains. Nat. Food 1, 437–446.

van Grinsven, H.J., ten Berge, H.F.M., Dalgaard, T., Fraters, B., Durand, P., Hart, A., Hofman, G., Jacobsen, B.H., Lalor, S.T.J., Lesschen, J.P., Osterburg, B., Richards, K.G., Techen, A.K., Vertès, F., Webb, J., Willems, W.J., 2012. Management, regulation and environmental impacts of nitrogen fertilization in northwestern Europe under the Nitrates Directive; a benchmark study. Biogeosciences 9, 5143–5160.

Vitousek, P.M., Naylor, R., Crews, T., David, M.B., Drinkwater, L.E., Holland, E., Johnes, P.J., Katzenberger, J., Martinelli, L.A., Matson, P.A., Nziguheba, G., Ojima, D., Palm, C.A., Robertson, G.P., Sanchez, P.A., Townsend, A.R., Zhang, F.S., 2009. Nutrient imbalances in agricultural development. Science 324 (5934).

Zhang, X., Davidson, E.A., Mauzerall, D.L., Searchinger, T.D., Dumas, P., Shen, Y., 2015. Managing nitrogen for sustainable development. Nature 528, 51–59.

Zingore, S., Murwira, H.K., Delve, R.J., Giller, K.E., 2007. Soil type, historical management and current resource allocation: three dimensions regulating variability of maize yields and nutrient use efficiencies on African smallholder farms. Field Crop. Res. 101, 296–305.

## Further reading

Bodirsky, B.L., Müller, C., 2014. Robust relationship between yields and nitrogen inputs indicates three ways to reduce nitrogen pollution. Environ. Res. Lett. 10, 111005.

Ferguson, R.B., 2015. Groundwater quality and nitrogen use efficiency in Nebraska's Central Platte River Valley. J. Environ. Qual. 44, 449–459.

Ladha, J.K., Jat, M.L., Stirling, C.M., Chakraborty, D., Pradhan, P., Krupnik, T.J., Sapkota, T.B., Pathak, H., Rana, D.S., Tesfaye, K., Gerard, B., 2020. Chapter Two - achieving the sustainable development goals in agriculture: the crucial role of nitrogen in cereal-based systems. In: Sparks, D.L. (Ed.), Advances in Agronomy. Academic Press, pp. 39–116.

Sanchez, P.A., 2002. Soil fertility and hunger in Africa. Science 295 (5562), 2019–2020.

van Grinsven, H.J., Bouwman, L., Cassman, K.G., van Es, H.M., McCrackin, M.L., Beusenet, A.H.W., 2015. Losses of ammonia and nitrate from agriculture and their effect on nitrogen recovery in the European Union and the United States between 1900 and 2050. J. Environ. Qual. 44, 356–367.

# Sources of nitrogen for crop growth: Pakistan's case

**Muhammad Aamer Maqsood[1], Naqsh-e-Zuhra[1], Imran Ashraf[1], Nasir Rasheed[1], Zia-ul-Hassan Shah[2]**

[1]*Institute of Soil and Environmental Sciences, University of Agriculture Faisalabad, Punjab, Pakistan;* [2]*Department of Soil Science, Sindh Agriculture University, Tandojam, Sindh, Pakistan*

## 2.1 Introduction

Nitrogen (N) is one of the most crucial nutrient elements which plants need to grow. Every fruit or grain harvested represents N extracted from soil (Sun et al., 2020). Nitrogen is plentiful in the earths' environment, as it makes up nearly 78% of the air we breathe. But the atmospheric nitrogen N is tightly bound that makes it unavailable to terrestrial plants (Wang et al., 2018). The plant available forms of N viz. ammonium ($NH_4^+$) and nitrate ($NO_3^-$) are relatively low in soil for plant use. Prior to the Green Revolution, farmers might not have known the complex chemical reactions of fertilizers in soil, but they were aware about the benefits of using animal manure, human waste, and compost application to soil. With increase in food demand, crop production also increased. To sustain high crop yield, organic fertilizer sources could not meet the crop nutrient requirements. Synthetic sources of fertilizers were introduced with urea as the main proponent to maintain high cop yields. It is now an accepted fact that urea and other conventional fertilizers have relatively low use efficiencies under conventional cropping systems. Application of commercial fertilizers in Pakistan began in 1952—53, and the offtake was only 1000 nutrient tonnes of N. During the early 1950s, fertilizer prices were highly subsidized to encourage the use of fertilizers among farmers (Dan and Bin, 2015). Nitrogenous fertilizers now account for almost 80% of commercial fertilizer off-take. The fertilizer demand is primarily driven by relative price, availability in the market, irrigation availability, acreage and cropping trend, etc. In Pakistan, a variety of fertilizers are in use and some of them are locally manufactured, while others are imported. Fertilizer recommendations for major crops are for nitrogen (N), phosphorus (P), and potassium (K). Most of the fertilizers are applied to irrigated wheat, cotton, sugarcane, maize, and rice crops. Over and unbalanced application of nitrogenous fertilizers is a problem in cotton and maize cropping regions. This over application not only results in loss of capital but also increases N pollution. Organic sources of N including farmyard manure (FYM), poultry manure, crop residues, sewage sludge, etc., are abundant in the country and there is need to cut

*Nitrogen Assessment. https://doi.org/10.1016/B978-0-12-824417-3.00005-8*

back on the application of synthetic N fertilizers and supplementation with organic fertilizer. This chapter highlights the types and amount of organic, synthetic, and non conventional sources of N available in Pakistan.

## 2.2 Organic sources of nitrogen

Soils of Pakistan are calcareous in nature with less than 1% organic matter (OM). Low soil OM is primarily due to high temperature and low rainfall in majority of the cropping regions. The sustainability of soil OM is dependent on soil physico-chemical properties, climatic conditions, type of organic material and its decomposition rate, and the rate at which native soil OM is mineralized. Soil OM is comprised of plant residues, microbial biomass, and humus. When residues of plant, animal remains, and microbial biomass fractions decompose, they release plant nutrients viz. nitrogen, phosphorus, potassium, etc. Therefore, organic fertilizers would naturally comprise of various plant-derived materials ranging from fresh or dried plant material to animal manures and agriculture industry by-products. The worldwide demand for organic foods production without synthetic inputs has contributed to the application of conservation practices, especially fertilization by means of organic waste (Trani et al., 2013). However, due to inherently low OM, farmers in Pakistan need to supply soil with high amounts of organic materials just to sustain a marginal level of OM for a short period of time.

### 2.2.1 Farmyard manure

Globally, livestock production has increased rapidly in the past century in response to increasing demand for livestock products (FAOSTAT, 2021b). In Pakistan, livestock sector contributes a major share to the value added products of agriculture ($\sim$56.3%) and nearly 11% to the agricultural gross domestic product (Rehman et al., 2017). Livestock production has also increased significantly during the past four decades in response to the increased food demand and changing dietary patterns in the country. As per Economic Survey of Pakistan 2019–20, total population of the livestock in the country is about 200 million animal heads (cattle, buffalo, sheep, and goats) (Government of Pakistan, 2019). As the livestock production has increased, the waste generation from the sector has also increased tremendously. This livestock waste is a primary constituent of FYM. In 2019, the world's total FYM (from cattle and buffalo source) application in agriculture was approximated at 12.46 mil tonnes. In Pakistan, during 2019, only 557,885 tonnes of FYM were applied to soil, which accounts for just 4.48% of FYM applied globally (FAOSTAT, 2021b).

FYM is a rich source of N and other nutrients, hence, can be utilized as organic fertilizer. Animal manure improves soil biological activity, improves nutrient cycling and availability for crops, and promotes soil aggregation. In comparison to synthetic fertilizers, regular application of FYM to soil also provides environmental benefits including increased soil carbon sequestration, reduction in water runoff and soil erosion, and lower $NO_3^-$ leaching. Empirical research has shown that regular application of FYM increases total soil N. Overall, improvement in soil health due to FYM application also improves crop growth and yield (Rayne and Aula, 2020).

### 2.2.2 Biological nitrogen fixation

Biological nitrogen fixation (BNF) is a fundamental part of the N cycle and accounts for a massive portion of the N available for plan uptake. The gaseous source of nitrogen ($N_2$) makes up 78% of atmospheric gases, but it is inert and inaccessible to plants. It cannot be utilized by the higher plants because the triple bond between the two atoms ($N = N$) of the atmospheric N can be broken only by some microorganisms, converting it into ammonia ($NH_3$) through nitrogenase enzymes ($N_2 + 8H^+ + 6e^- = 2NH_3 + H_2$) (Newton, 2015; Chanway et al., 2014). This fixation of N by microbes is termed as BNF and contributes a significant share in N availability to crops. Globally, $52-130\,Tg\,N\,year^{-1}$ is fixed through BNF (Davies-Barnard and Friedlingstein, 2020). The N fixer microbes can live freely or in association with certain plant species. Symbiotic nitrogen fixation is responsible for adding $20-200\,kg\,N\,ha^{-1}year^{-1}$. The symbiotic formation of nodules by Rhizobia on the roots of legume crops is a highly specialized and efficient processes for N2 fixation that also substitutes the soil N pool (Ladha and Reddy, 2000). As a legume, groundnut/peanut (*Arachis hypogea*) is also known to fix at around $50-100\,kg\,N\,ha^{-1}\,year^{-1}$ (Oteng-Frimpong and Dakora, 2018). The major groundnut producing districts of Pakistan include Chakwal, Attock, and Rawalpindi in Punjab, Karak, and Sawabi in Khyber Pakhtunkhwa (KPK), and Sanghar in Sindh province (Khalid et al., 2020). Different varieties of groundnut are grown on an area about 82.9 thousand hectares with an average yield of $0.61\,tonnes\,ha^{-1}$ (Qasim et al., 2016). Similarly, mung-bean grown on 127 thousand hectares (Government of Pakistan, 2015) can fix approx. $47\,kg\,N\,ha^{-1}$ during a cropping season (Hayat et al., 2008). Kallar grass (*Laptochloa fusca*) also a non conventional fodder crop in Pakistan may fix $15-32\,kg\,N\,ha^{-1}$ (Malik et al., 1988).

### 2.2.3 Green manuring

Growing crops, preferably legumes, and incorporating them into soil when they achieve an optimum output of green tops is referred to as green manuring. Since legumes can fix N from the atmosphere, these are favored for green manuring which improves N contents and soil OM. A constant use of green manure improves OM contents and soil's nutrient pool, enhancing the soil's physical, chemical, and biological properties (Das et al., 2020). On a global scale approximately $100-120$ million hectares of land is cultivated for forage and green manuring (Smil, 1999). Legume crops such as dhaincha, sesbania sunn hemp, cowpeas, and guar (during summer) and berseem and senji (during winter) have shown promising results for improving soil health and addition of significant amounts of N to soil. According to Tanveer et al. (2019), these legumes add 45 to $118\,kg\,N\,ha^{-1}$ depending upon their species, soil properties, and climatic conditions.

### 2.2.4 Crop residues

Crop residue addition to soil is a good source of OM that conserves the carbon stock in soil, helps to maintain soil moisture, recycles nutrients, and reduces the environmental hazards by lowering the amount of residue burning in fields (Liang et al., 2016). Globally, cereals account for 74% of crop residue production, 10% contributed by sugar crops, 8% by legumes, 5% by tubers, and just 3% by oilseeds crops (Lal, 2005). Depending upon the soil fertility and crop species, after decomposition, crop residue contributes several mineral nutrients (Lal, 2005). However, crop residue with high C:N ratio may initially immobilize the available soil N (Garai et al., 2020). Nonetheless, continuous and long-term addition of crop residues ultimately leads of improved nutrient availability for subsequent crops.

A huge amount of crop residues is produced annually in Pakistan. The return of crop residues into the soil is a common way to preserve soil OM, boost ecological activities, physical properties, and recycling of nutrients, especially N (Smitha et al., 2019). According to Khalil et al. (2005), addition of 1% (on dry soil basis) may release up to 141 kg N ha$^{-1}$ in calcareous soils. Although adding such huge amounts of crop residues during a cropping year may be unrealistic, however, progressive incorporation of crop residues over several years may moderately reduce reliance on synthetic N fertilizers. On average, 1.44 tonnes acre$^{-1}$ and 1.32 tonnes acre$^{-1}$ of residues are produced from rice and wheat crop, respectively (Rafiq et al., 2019). During 2017−18, 5.9 million tonnes of corn stover, 7.45 million tonnes of rice husk, 25.08 million tonnes of wheat straw, and 1.68 million tonnes of cotton sticks were produced as crop residues (Kashif et al., 2020). These residues contain, on average, 0.8%, 0.65%, 0.75%, and 1% N on dry weight basis for maize, rice, wheat, and cotton, respectively (Kashif et al., 2020).

### 2.2.5 Poultry waste

Poultry production is an important agriculture-based industry in Pakistan. During 2019, the estimated number of chickens, commercially raised, was 1.321 billion (FAOSTAT, 2021a,b). The poultry industry produces a high amount of waste, i.e., 5 kg chicken$^{-1}$ year$^{-1}$ (Chastain et al., 2001). In Pakistan, during 2019, only 268,393 tonnes of poultry waste/manure were applied to soil, which accounts for just 4.3% of poultry waste applied globally (FAO, 2021). Poultry manure is a significant source of plant nutrients. One tonne of dry poultry litter contains 26 kg N, 18 kg P, and 10 kg K (Kyakuwaire et al., 2019).

### 2.2.6 Sewage sludge

Increasing human population has put tremendous pressure on freshwater resources. This increase in not only depleting the finite amount of freshwater but also creating huge amounts of wastewater. Pakistan produced 2.48 million acre feet of wastewater in 2017 (FAO, 2021). The amount of wastewater that is treated for agricultural use is not documented. However, all major cities of Pakistan have sewage treatment plants.

Sewage sludge is the residual, semisolid material that is produced as a by-product during sewage treatment of municipal wastewater. Even though there have been advancements in techniques to remove nutrients from wastewater, the nutrient discharge to surface water is still projected to increase up to 70% by the year 2050 in developing countries. Nitrogen content in this discharge may significantly increase from 10.4 Tg in 2010 to 13.5−17.9 Tg, by 2050. Nutrient discharge into sewage water can be reduced by using a separate collection system for urine that can globally save 15 Tg N year$^{-1}$ for use in agriculture (Van Puijenbroek et al., 2019).

Sewage sludge, as mentioned, is produced in bulk quantities and is a good source of OM. Sewage sludge contains 20%−30% proteins and corresponding 1.5%−4% nitrogen (Kominko et al., 2017). Therefore, sewage sludge application to soil as a supplement of N fertilizer may be a viable disposal option for this waste product. Systematic collection of sewage sludge for use in agriculture application is not present in Pakistan and thus the use of sewage sludge as an organic supplement to soil is not a common practice. However, with increasing environmental concerns of unorganized dumping of sewage sludge and the realization that sewage sludge could help increase agricultural productivity, the situation may change in the near future. Local research has revealed beneficial effects of sewage sludge application on crop production and improvement in soil properties (Azam et al., 1999; Murtaza et al., 2010; Riaz et al., 2018). However, the positive effects of sludge are not persistent due to the high rate of decomposition.

## 2.3 Synthetic sources of nitrogen for crops

During the past five decades, the use of synthetic fertilizers in Pakistan has been substantially increased (Rehman et al., 2019). Nitrogenous fertilizers were introduced in Pakistan in 1952, phosphorus 7 years later in 1959/60, and potassium another 7 years later in 1966/67. fertilizer use gained momentum after 1966/67, when high yielding varieties of cereal crops were introduced. In Asia, Pakistan is among the largest producing countries of synthetic N fertilizer after China and India. Urea, diamonium phosphate, nitrophos, calcium amonium nitrate, sulfate of potash, and single super phosphate and NPK are typical locally produced fertilizers either on large or small scale (Table 2.1).

The local production of urea was largest among all other fertilizers of Pakistan during 2017−18 and 2018−19. The production of urea was increased compared to the year 2017−18 and remained 75.8% in 2018−19 followed by diammonium phosphate (DAP) 10% of the total production of locally produced fertilizers in the year 2018−19. The share of NP and CAN remained 6.3% and 5.8%. Other fertilizer production remained only 2.1% in total production of fertilizer during the year 2018−19 (Annual Fertilizer Review, 2018−19) (Fig. 2.1).

According to a national fertilizer use survey five major crops (wheat, cotton, sugarcane, rice, and maize) account for 87% of fertilizer consumption. Overall, Pakistan consumes about 3% of the total N used in the world (Raza et al., 2018).

**Table 2.1** List of fertilizer manufacturing companies of Pakistan and their products.

| Sr # | Company name | Products |
|------|--------------|----------|
| 1 | Fauji Fertilizer Company Limited (FFC) | Urea, DAP, SOP, MOP |
| 2 | Fauji Fertilizer Bin Qasim Limited (FFBL) | Urea |
| 3 | Engro Fertilizers Limited (EFL) | Urea, NPK, NP |
| 4 | Dawood Hercules Chemicals Limited (DHCL) | Urea |
| 5 | Pak Arab Fertilizers Limited (PFL) | Urea, NP, CAN |
| 6 | Fatima Fertilizers Company Limited | Urea, CAN, NP, NPK |
| 7 | Agritech Limited (formerly Pak American Fertilizers Limited) | Urea |
| 8 | Engro Chemicals Pakistan Limited (ECPL) | NP |
| 9 | Lyallpur Chemicals & Fertilizers Limited (LC & FL) | SSP |
| 10 | Hazara Fertilizer | SSP |
| 11 | Suraj Fertilizer | SSP |

*Reproduced from Fertilizer Review. 2018–19. National Fertilizer Development Center. Government of Pakistan, Islamabad. http://www.nfdc.gov.pk/public.html (Pakistan Bureau of Statistics).*

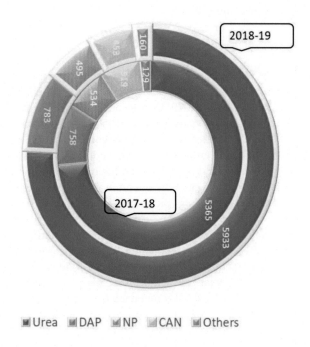

**FIGURE 2.1**

Domestic production of fertilizer (000 tonnes) during 2017–18 and 2018–19.

*Modified from Fertilizer Review. 2017–18. National Fertilizer Development Center. Government of Pakistan, Islamabad. http://www.nfdc.gov.pk/public.html; Fertilizer Review. 2018–19. National Fertilizer Development Center. Government of Pakistan, Islamabad. http://www.nfdc.gov.pk/public.html (Pakistan Bureau of Statistics).*

### 2.3.1 Urea

French chemist Hilaire-Marin Rouelle isolated urea for the first time from urine in 1773. It's preparation was done from ammonium cyanate by the German chemist Friedrich Wöhler in 1828. This was the first generally accepted organic compound in a laboratory which was synthesized for an inorganic compound (Janssen, 2021). Urea is now being prepared on a commercial scale, in large amounts from liquid carbon dioxide and liquid ammonia. These two materials are combined under elevated temperatures and high pressures for the synthesis of ammonium carbamate, decompose at much lower pressures, and produce urea along with water.

Urea is the most important and commonly used N-based fertilizer across the world and shares 55% to the total inorganic fertilizer market (Heffer and Prud-homme, 2015). In 2019, the world total urea consumption in agriculture was approximated at 65.2 mil tonnes. The leading five countries that account for 82.63% of urea use include India, United States, Pakistan, Indonesia, and Canada. Pakistan's share in agricultural use of urea is around 9.55% (FAO, 2021). In Pakistan, urea is also a major source of N fertilizer, with a share of 72% and 85% during the years 1981−85 and 2009−13, respectively. The production of urea in Pakistan progressed due to the installation of various plants of urea formulation to meet the country's demand. The approximate share of urea ranged from 72% to 85% in total N-based fertilizers consumed in Pakistan during 1981−2013. From 2008 to 2019, Pakistan saw a 13.7% increase in the use of urea in Pakistan (FAO, 2021).

Though urea production in Pakistan changed substantially in the recent years, it showed an overall increasing trend through 2003−2017. The domestic production of urea in 2017−18 was 5.65 million tonnes which increased to 5.93 million tonnes during 2018−19.

An increase of 32.46% in locally produced urea fertilizer was recorded in 2016 as compared to the production in 2003 (Fig. 2.2). The import of urea in early years (2003, 2004) was not remarkable but, later, urea was imported on larger scale to meet the demand of intensive agriculture. The maximum import of urea (1.231 million tonnes) was made in 2011. In the recent years, due to the increase in local production of urea, the import has been decreased significantly.

The largest market share of urea was 29% by Fauji Fertilizer Company. The share of urea by other local manufacturing companies is given in Fig. 2.3.

### 2.3.2 Diammonium phosphate

DAP is the world's most widely used phosphorus fertilizers that also contains N (18%). In 2019, the world total DAP consumption in agriculture was approximated at 17.2 mil tonnes. India is the largest agricultural consumer of DAP in the world. The leading five countries that account for 92.01% of DAP use include India, United States, Pakistan, Bangladesh, and Turkey. Pakistan's share in agricultural use of urea is around 11.81% (FAO, 2021).

The ammonium present in DAP is an excellent N source. The initial pH of DAP solution is basic and can influence the microsite reactions of phosphate and soil OM

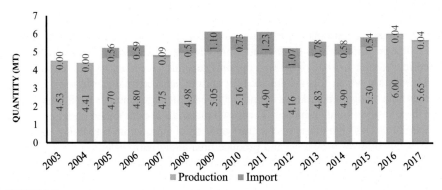

**FIGURE 2.2**

Local Production and Import of Urea in Pakistan for the period 2003–17 (FAOSTAT, 2018).

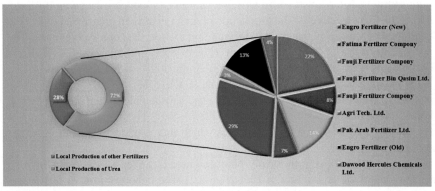

**FIGURE 2.3**

Share of manufacturing companies in local production of urea.

*Modified from Fertilizer Review. 2018–19. National Fertilizer Development Center. Government of Pakistan, Islamabad. http://www.nfdc.gov.pk/public.html (Pakistan Bureau of Statistics).*

(Zhang et al., 2017). However, nitrification of ammonium ($NH_4^+$) and its uptake by plants results in a subsequent drop in pH. Therefore, the rise in soil pH surrounding DAP granules is a temporary effect. A minor portion of DAP fertilizer is being produced locally and the rest of it is imported to meet the demand of this fertilizer. The only manufacturer of this fertilizer is Fauji Fertilizer Company Ltd.

The demand and utilization of DAP fertilizer increased in Pakistan gradually, especially in the last decade. Major portion of this fertilizer is imported without substantial increase in local production capacity. Fig. 2.4 shows the local production and import of this fertilizer from 2003 to 2017. The maximum availability of DAP was in 2017 in which 809 1000 tonnes was produced locally while 1730 1000 tonnes was imported.

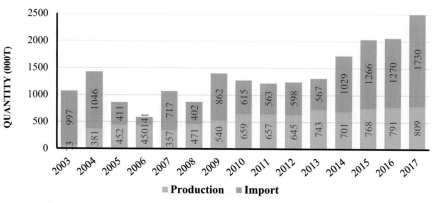

**FIGURE 2.4**

Local Production and Import of DAP in Pakistan during the period 2003—17 (FAOSTAT, 2018).

### 2.3.3 Nitrophos

Nitrophos (NP) is a granulated fertilizer being produced in Pakistan that has equal amounts of phosphorus (P) and nitrogen (N). The grade of this fertilizer varies depending upon the combination of nutrient used in it. In addition to this, NP being a highly acidic product with a pH of 3.5 is the most suitable fertilizer for soils that have a high pH and are alkaline in nature.

The production and application of NP fertilizers is largely regional. The production process uses nitric acid instead of sulfuric acid for treating phosphate rock and does not produce gypsum by-products. The balanced combination of nitrogen and phosphorus is ideal for plant growth and development.

In many crops, it could be effectively applied at planting and at early growth stages of the crops. Each grain has equal amounts of N and P; hence, the nutrients are equally distributed throughout the field. Locally produced NP contains 23% N and 23% $P_2O_5$. Half of the N is in ammoniacal form and the other half is in nitrate form. The water solubility of phosphorus in 23-23-0 grade is normally more than 70%.

Pakistan is manufacturing most of NP fertilizer locally and very little quantity was being imported. An increasing trend of local production has been observed in last few years. Maximum local production (704 thousand tonnes) and import (30.3 thousand tonnes) was recorded in 2015 (Fig. 2.5).

### 2.3.4 Calcium ammonium nitrate

Calcium ammonium nitrate (CAN) contains 26—28% N and 8—10% calcium. It is commercially produced by combining ammonium nitrate and calcium carbonate in a 3:1 ratio. Half of the N is in ammoniacal form and half in nitrate form. The major use of CAN is in fruit and vegetables crops, as the additional calcium is good for rooting, stress-free growth, strong cell walls, improved fruit quality, and better storage. CAN is a good alternative of urea for most crops but has a lower efficiency compared to other commercial sources of N for rice crop because a major part of

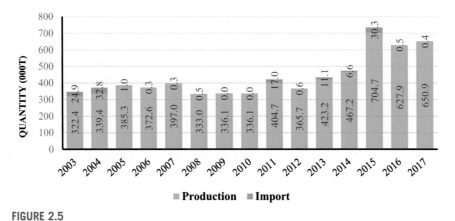

**FIGURE 2.5**

Local Production and Import of NP in Pakistan during the period 2003–17 (FAOSTAT, 2018).

the nitrate component may be denitrified and lost under flooder soil conditions. The leading five countries that account for 65.60% global use of CAN include Turkey, Hungary, Czechia, Pakistan, and Ireland. Pakistan's share in agricultural use of CAN is approximately 9.44%. In Pakistan, CAN is ranked third in agricultural use after urea and DAP. The production and availability (especially in KPK province) of CAN has faced many hurdles in the past due to ammonium nitrate use as a potential explosive. The agricultural use of CAN progressively increased till 2017 reaching 0.731 million tonnes but reduced by 35% in 2019 (FAOSTAT, 2021a). The government subsidy on urea and a decrease in local production of CAN are partially responsible for decline in use.

## 2.4 Non conventional sources of N

The NUE in Pakistan is in the range of 30%–50% (Raza et al., 2018). The situation can be further worsened by changing climatic conditions and poor management practices. Extensive research has been conducted on management practices and technologies that can substantially improve NUE and reduce environmental pollution (Choudhary et al., 2018; Keuschnig et al., 2020). Slow release fertilizers are a 21st-century advancement in improving NUE. As classified by the Association of American Plant Food Control Officials (AAPFCO), slow release fertilizers delay the release of plant nutrient or extend its availability to the plant/crop for a significantly longer time as compared to conventional fertilizers such as urea, ammonium nitrate, etc. (Fu et al., 2018). One of the strategies for reducing N losses and increasing efficiency and yields of major crops is the use of nitrification inhibitors (NIs) (Dawar et al., 2021). NIs delay bacterial conversion of the $NH_4^+$ by *Nitrosomonas* bacteria and thus sustain $NH_4^+$ for longer period in soil for plant uptake. NIs can provide greater opportunity for synchronizing N release with crop demand (Singh, 2010), potentially reducing $N_2O$ emission and $NO_3^-$ leaching.

**Table 2.2** Use of Natural Nitrification inhibitors in Pakistan.

| Natural nitrification inhibitors | Crop/ soil | References |
|---|---|---|
| Neem oil—coated urea | Maize | Ali et al. (2020) |
| Neem oil—coated urea, moringa oil—coated urea, and pomegranate leaves extract—coated urea | Maize | Ashraf et al. (2019) |
| Karanj, Dharek, and calcium carbide | Wheat | Abbasi and Manzoor (2013) |

Application of NIs is found as beneficial approach to minimize N losses and for limiting environmental threats caused by nitrogenous fertilizers (Prasad, 2009). There are many plant products or by-products which are being used as NIs including neem (*Azadirachta indica*), Karanj (*Pongomia glabra*), vegetable tannins, waste products of tea, mint oil, Japanese mint, and Mustard (*Brassica juncea* L.), having strong antibacterial activities and are being used in agriculture to increase the NUE (Table 2.2). Some of the natural products such as neem have ability to inhibit urease activity as well (Patra et al., 2006).

Many synthetic NIs have been introduced which block the path of enzyme, particularly ammonia monooxygenase (AMO), which is responsible for oxidation of ammonium to nitrate. Ammonia monooxygenase (AMO) can be inhibited by the compounds having thiono-S and bind to the Cu in their active sites, while some inactivate the AMO by their heterocyclic ring of N (McCarty, 1999). The three majors synthetic NIs are Dicyandiamide (DCD), Nitrapyrin, and 3, 4-Dimethyl pyrazol phosphate (DMPP) (Subbarao et al., 2006). DCD application rates of 16—20 g/kg urea or 1.4—1.8 kg DCD/ha are reported, while DMPP is applied at 10 g/kg urea, or 0.5—1.5 kg DMPP/ha (Zerulla et al., 2001; Harty et al., 2016). Collectively, these NIs can reduce $N_2O$ emission by as much as 38%, on average (Akiyama et al., 2010). NIs that have been widely evaluated in Pakistan are given in Table 2.3.

The inhibition of urease and nitrification activities in the N cycle is considered as one of the best approaches to reduce N losses and increase its use efficiency.

**Table 2.3** Use of synthetic nitrification inhibitors in Pakistan.

| Synthetic nitrification inhibitors | Crop/soil | References |
|---|---|---|
| Nitrapyrin | Wheat | Dawar et al. (2021) |
| DCD | Pastoral systems | Zaman et al. (2013) |
| Super urea (NBPT and DCD) | Wheat | Khan et al. (2013) |
| 1H-benzotriazole; 4-amino-1,2,4-triazole; benzothiazole; 3-methylpyrazole-1-carboxamide; 4-bromo-3-methylpyrazole; pyrazole; lignosulfonic acid | Soil | Ali et al. (2012) |

Hydrolysis of urea by urease enzyme releases ammonium ion in soil which can be volatilized as ammonia under high temperature in Pakistan, as urease activity increases up to 40°C. Accordingly, central to enhanced efficiency fertilizers (EEFs) has been the use of inhibitors to modify fertilizer release patterns by altering the biological activities of N-metabolizing enzymes of soil bacteria (Chien et al., 2009; Timilsena et al., 2015). Urease inhibitors are formulated into solid (e.g., by coating) or liquid fertilizer products, or applied separately to the soil primarily to reduce urea hydrolysis by ureases. The most widely used urease inhibitor is N-(n-Butyl) thiophosphoric triamide (NBPT). NPBT application has been reported to reduce up to 66% $NH_3$ losses from urea (Panel, 2015). Other common urease inhibitors include cyclohexyl phosphoric triamide (CHPT), ammonium thiosulfate (ATS), phenylphosphorodiamide/phenylphosphorodiamidate (PPD/PPDA), hydroquinone, and calcium thiosulfate (CTS).

Another approach to decrease volatilization loss is coating of urea with polymeric substances which decrease dissolution rates. Polymer-coated fertilizers (PCFs) are categorized as technologically most efficient and advanced EEFs due to their nutrient delivery system. These may be slow or controlled release fertilizers; thus, extending the availability of N for longer period of time. In PCFs, the fertilizer core is covered by protective barrier of polymer which is resistant to rapid hydrolysis and microbial degradation (Ellison et al., 2013). The polymer coating has micropores that allow soil moisture to diffuse through the coating to dissolve the fertilizer and allow it to eventually reach the soil solution (Adams et al., 2013).

Commercial availability and farmer use of non conventional, high efficiency fertilizers in Pakistan is negligible. However, research on formulation and crop response in ongoing. Saleem et al. (2021) reported that availability of nitrogen can be prolonged up to 60 days when conventional DAP is coated with $KFeO_2$ nanoparticles.

## 2.5 Conclusion

Pakistan is a developing country, with an agriculture-based economy. Crop production primarily relies on synthetic fertilizers with heavy reliance on urea. Highest amount of N fertilizer produced in Pakistan is also urea. Even though a substantial amount of N may be added to soil via organic sources. Still, organic N is a neglected and least documented aspect of agriculture in Pakistan. Most of the information collected and published by the national institutes linked to the agriculture sector is focused on synthetic fertilizer production and consumption. There needs to be an extensive survey on the extent and type of organic fertilizers/amendment used at farm level. Furthermore, low fertilizer efficiency and significant environmental losses calls for a shift to EEF. Encouraging research has been published on local synthesis and beneficial use of urea/DAP coated with natural NIs. However, commercial production and farm level use is not present. The government bodies need to encourage extensive research and promote the industrial production and farmer adoption of new and efficient fertilizers.

# References

Abbasi, M.K., Manzoor, M., 2013. Effect of soil-applied calcium carbide and plant derivatives on nitrification inhibition and plant growth promotion. Int. J. Environ. Sci. Technol. 10, 961–972.

Adams, C., Frantz, J., Bugbee, B., 2013. Macro-and micronutrient-release characteristics of three polymer-coated fertilizers: theory and measurements. J. Plant Nutr. Soil Sci. 176 (1), 76–88.

Akiyama, H., Yan, X.Y., Yagi, K., 2010. Evaluation of effectiveness of enhanced-efficiency fertilizers as mitigation options for $N_2O$ and NO emissions from agricultural soils: meta-analysis. Glob. Change Biol. 16, 1837–1846.

Ali, R., Kanwal, H., Iqbal, Z., Yaqub, M., Khan, J.A., Mahmood, T., 2012. Evaluation of some nitrification inhibitors at different temperatures under laboratory conditions. Soil Environ. 31, 134–145.

Ali, M., Maqsood, M.A., Aziz, T., Awan, M.I., 2020. Neem (*Azadirachta indica*) oil coated urea improves nitrogen use efficiency and maize growth in an alkaline calcareous soil. Pak. J. Agric. Sci. 57, 675–684.

Ashraf, M.N., Aziz, T., Maqsood, M.A., Bilal, H.M., Raza, S., Zia, M., Mustafa, A., Xu, M., Wang, Y., 2019. Evaluating organic materials coating on urea as potential nitrification inhibitors for enhanced nitrogen recovery and growth of maize (Zea mays). Int. J. Agric. Biol. 22, 1102–1108.

Azam, F., Ashraf, M., Lodhi, A., Gulnaz, A., 1999. Utilization of sewage sludge for enhancing agricultural productivity. Pak. J. Bio. Sci. 2, 370–377.

Chanway, C.P., Anand, R., Yang, H., 2014. Nitrogen fixation outside and inside plant tissues (pp 3–2). In: Advances in Biology and Ecology of Nitrogen Fixation. InTech, Croatia. https://doi.org/10.5772/57532.

Chastain, J.P., Camberato, J.J., Skewes, P., 2001. Poultry manure production and nutrient content (pp 1–17). In: Confined Animal Manure Managers Certification Program Manual: Poultry Version. Clemson University Extension, Clemson SC, USA.

Chien, S.H., Prochnow, L.I., Cantarella, H., 2009. Recent developments of fertilizer production and use to improve nutrient efficiency and minimize environmental impacts. Adv. Agron. 102, 267–322.

Choudhary, M., Panday, S.C., Meena, V.S., Singh, S., Yadav, R.P., Mahanta, D., Mondal, T., Mishra, P.K., Bisht, J.K., Pattanayak, A., 2018. Long-term effects of organic manure and inorganic fertilization on sustainability and chemical soil quality indicators of soybean-wheat cropping system in the Indian mid-Himalayas. Agric. Ecosys. Environ. 257, 38–46.

Dan, A.A.C.J.Y., Bin, L.X.G.T., 2015. Time series analysis of the impact of rising prices of inorganic fertilizers on field crops production: a case study of Pakistan. J. Econ. Sustain. Dev. 6, 62–71.

Das, K., Biswakarma, N., Zhiipao, R., Kumar, A., Ghasal, P.C., Pooniya, V., 2020. Significance and management of green manures. In: Soil Health. Springer, Cham, pp. 197–217. https://doi.org/10.1007/978-3-030-44364-1_12.

Dawar, K., Khan, A., Sardar, K., Fahad, S., Saud, S., Datta, R., Danish, S., 2021. Effects of the nitrification inhibitor nitrapyrin and mulch on $N_2O$ emission and fertilizer use efficiency using 15N tracing techniques. Sci. Total Environ. 757, 143739.

Davies-Barnard, T., Friedlingstein, P., 2020. The global distribution of biological nitrogen fixation in terrestrial natural ecosystems. Global Biogeochem. Cycles 34. https://doi.org/10.1029/2019GB006387 e2019GB006387.

Ellison, E., Blaylock, A., Sanchez, C., Smith, R., 2013. Exploring controlled release nitrogen fertilizers for vegetable and melon crop production in California and Arizona. In: Western Nutrient Management Conference. Reno, NV. USA, pp. 17−22.

FAOSTAT, 2021a. Food and Agriculture Organization. Fertilizer by Product. http://www.fao.org/faostat/en/#data/RFB.

FAOSTAT, 2021b. Food and Agriculture Organization. Livestock Manure. http://www.fao.org/faostat/en/#data/EMN.

FAOSTAT, 2018. Food and Agriculture Organization. Fertilizer by Product. http://www.fao.org/faostat/en/#data/RFB.

FAO, 2021. AQUASTAT Database. http://www.fao.org/nr/water/aquastat/data/query/index.html. (Accessed 29 March 2021).

Fertilizer Review, 2018-19. National Fertilizer Development Center. Government of Pakistan, Islamabad. http://www.nfdc.gov.pk/public.html.

Fu, J., Wang, C., Chen, X., Huang, Z., Chen, D., 2018. Classification research and types of slow controlled release fertilizers (SRFs) used - a review. Commun. Soil Sci. Plant Anal. 49, 2219−2230. https://doi.org/10.1080/00103624.2018.1499757.

Garai, S., Mondal, M., Mukherjee, S., 2020. In: Maitra, S., Pramanick, B. (Eds.), Smart Practices and Adaptive Technologies for Climate Resilient Agriculture. New Delhi Publishers, Kolkata, India, pp. 327−358.

Government of Pakistan, 2015. Pakistan Economic Survey 2014-15. Available online: at: https://www.finance.gov.pk/survey_1415.html

Government of Pakistan, 2019. Pakistan Economic Survey 2017-18. Available online at: http://www.finance.gov.pk/survey_1718.html.

Hayat, R., Ali, S., Ijaz, S.S., Chatha, T.H., Siddique, M.T., 2008. Estimation of $N_2$ fixation of mung bean and mash bean through xylem ureide technique under rainfed conditions. Pak. J. Bot. 40, 723−734.

Harty, M.A., Forrestal, P.J., Watson, C.J., McGeough, K.L., Carolan, R., Elliot, C., Krol, D., Laughlin, R.J., Richards, K.G., Lanigan, G.J., 2016. Reducing nitrous oxide emissions by changing N fertilizer use from calcium ammonium nitrate (CAN) to urea-based formulations. Sci. Total Environ. 563−564, 576−586.

Heffer, P., Prud'homme, M., May 2015. Fertilizer outlook 2015−2019. In: 83rd IFA Annual Conference, vol. 415. International Fertilizer Industry Association (IFA), Istanbul (Turkey).

Kashif, M., Awan, M.B., Nawaz, S., Amjad, M., Talib, B., Farooq, M., Nizami, A.S., Rehan, M., 2020. Untapped renewable energy potential of crop residues in Pakistan: Challenges and future directions. J. Environ. Manag. 256, 109924.

Keuschnig, C., Gorfer, M., Li, G., Mania, D., Frostegård, Å., Bakken, L., Larose, C., 2020. NO and $N_2O$ transformations of diverse fungi in hypoxia: evidence for anaerobic respiration only in Fusarium strains. Environ. Microbiol. 22, 2182−2195.

Khalid, R., Zhang, X., Hayat, R., Ahmed, M., 2020. Molecular characteristics of Rhizobia isolated from *Arachis hypogaea* grown under stress environment. Sustainability 12, 6259.

Khalil, M., Hossain, M., Schmidhalter, U., 2005. Carbon and nitrogen mineralization in different upland soils of the subtropics treated with organic materials. Soil Biol. Biochem. 37, 1507−1518.

Khan, M.A., Shah, Z., Rab, A., Arif, M., Shah, T., 2013. Effect of urease and nitrification inhibitors on wheat yield. Sarhad J. Agric. 29, 371−378.

Kominko, H., Gorazda, K., Wzorek, Z., 2017. The possibility of organo-mineral fertilizer production from sewage sludge. Waste Biomass Valor. 8, 1781−1791. https://doi.org/10.1007/s12649-016-9805-9.

Kyakuwaire, M., Olupot, G., Amoding, A., Nkedi-Kizza, P., Basamba, T.A., 2019. How Safe is chicken litter for land application as an organic fertilizer? A review. Int. J. Environ. Res. Public Health 16 (19), 3521. https://doi.org/10.3390/ijerph16193521.

Ladha, J.K., Reddy, P.M., 2000. The quest for nitrogen fixation in rice. In: Proceedings of the 3rd Working Group Meeting on Assessing Opportunities for Nitrogen Fixation in Rice. International Rice Research Institute, Makati, p. 354.

Lal, R., 2005. World crop residues production and implications of its use as a biofuel. Environ. Int. 31, 575–584.

Liang, F., Li, J., Yang, X., Huang, S., Cai, Z., Gao, H., Ma, J., Cui, X., Xu, M., 2016. Three-decade long fertilization-induced soil organic carbon sequestration depends on edaphic characteristics in six typical croplands. Sci. Rep. 6, 30350. https://doi.org/10.1038/srep30350.

Malik, K.A., Bilal, R., Azam, F., Sajjad, M.I., 1988. Quantification of $N_2$-fixation and survival of inoculated diazotrophs associated with roots of Kallar grass. Plant Soil 108, 43–51.

McCarty, G.W., 1999. Modes of action of nitrification inhibitors. Biol. Fert. Soils 29, 1–9.

Murtaza, G., Ghafoor, A., Qadir, M., Owens, G., Aziz, M.A., Zia, M.H., Saifullah., 2010. Disposal and use of sewage on agricultural lands in Pakistan: a review. Pedosphere 20, 23–34.

Newton, W.E., 2015. Recent advances in understanding nitrogenases and how they work. In: de Bruijn, F.J. (Ed.), Biological Nitrogen Fixation. https://doi.org/10.1002/9781119053095.ch2.

Oteng-Frimpong, R., Dakora, F.D., 2018. Selecting elite groundnut (*Arachis hypogaea* L) genotypes for symbiotic N nutrition, water-use efficiency and pod yield at three field sites, using 15N and 13C natural abundance. Symbiosis 75, 229–243. https://doi.org/10.1007/s13199-017-0524-1.

Panel, E.N.E., 2015. Nitrogen Use Efficiency (NUE) an Indicator for the Utilization of Nitrogen in Food Systems. Wageningen University, Alterra: Wageningen, The Netherlands.

Patra, D.D., Kiran, U., Pande, P., 2006. Urease and nitrification retardation properties in natural essential oils and their by-products. Commun. Soil Sci. Plant Anal. 37 (11–12), 1663–1673.

Prasad, R., 2009. Efficient fertilizer use: the key to food security and better environment. J. Trop. Agric. 47 (1), 1–17.

Qasim, M., Bakhsh, K., Tariq, K.A., Nasir, M., Saeed, R., Mahmood, M.A., 2016. Factors affecting groundnut yield in Pothwar region of Punjab, Pakistan. Pak. J. Agric. Res. 29, 76–83.

Raza, S., Zhou, J., Aziz, T., Afzal, M.R., Ahmed, M., Javaid, S., Chen, Z., 2018. Piling up reactive nitrogen and declining nitrogen use efficiency in Pakistan: a challenge not challenged (1961–2013). Environ. Res. Lett. 13 (3), 034012.

Rehman, A., Chandio, A.A., Hussain, I., Jingdong, L., 2019. Fertilizer consumption, water availability and credit distribution: major factors affecting agricultural productivity in Pakistan. J. Saudi Soc. Agric. Sci. 18 (3), 269–274.

Rehman, A., Jingdong, L., Chandio, A.A., Hussain, I., 2017. Livestock production and population census in Pakistan: determining their relationship with agricultural GDP using econometric analysis. Inf. Process. Agric. 4, 168–177. https://doi.org/10.1016/j.inpa.2017.03.002.

Rafiq, M., Ahmad, F., Atiq, M., 2019. The determinants of the crop residue management in Pakistan: an environmental appraisal. Bus. Econ. Rev. 11, 179–200.

Rayne, N., Aula, L., 2020. Livestock manure and the impacts on soil health: a review. Soil Syst. 4, 64. https://doi.org/10.3390/soilsystems4040064.

Riaz, U., Murtaza, G., Ullah, S., Farooq, M., 2018. Influence of different sewage sludges and composts on growth, yield, and trace elements accumulation in rice and wheat. Land Degrad. Dev. 1−10. https://doi.org/10.1002/ldr.2925.

Saleem, I., Maqsood, M.A., Rehman, M.Z., Aziz, T., Bhatti, I.A., Ali, S., 2021. Potassium ferrite nanoparticles on DAP to formulate slow release fertilizer with auxiliary nutrients. Ecotoxic. Environ. Safety. https://doi.org/10.1016/j.ecoenv.2021.112148.

Singh, U., 2010. Nutrient efficient fertilizers and practices relating to food security and sustainability. In: TFI/FIRT Fertilizer Outlook and Technology Conference, November 16−18, Savannah, GA USA.

Smil, V., 1999. Nitrogen in crop production: an account of global flows. Global Biogeochem. Cycles 13 (2), 647−662. https://doi.org/10.1029/1999GB900015.

Smitha, G.R., Basak, B.B., Thondaiman, V., Saha, A., 2019. Nutrient management through organics, bio-fertilizers and crop residues improves growth, yield and quality of sacred basil (*Ocimum sanctum* Linn). Indus. Crops Prod. 128, 599−606.

Subbarao, G.V., Ito, O., Sahrawat, K.L., Berry, W.L., Nakahara, K., Ishikawa, T., Watanabe, T., Suenaga, K., Rondon, M., Rao, I.M., 2006. Scope and strategies for regulation of nitrification in agricultural system: Challenges and opportunities. Crit. Rev. Plant Sci. 25, 303−335.

Sun, Y., Wang, M., Mur, L.A.J., Shen, Q., Guo, S., 2020. Unravelling the roles of nitrogen nutrition in plant disease defences. Int. J. Mol. Sci. 21 (2), 572.

Tanveer, A., Ali, H.H., Ikram, N.A., 2019. Green manuring for soil health and sustainable production of agronomic crops. In: Hasanuzzaman, M. (Ed.), Agronomic Crops. Springer, Singapore. https://doi.org/10.1007/978-981-32-9783-8_20.

Timilsena, Y.P., Adhikari, R., Casey, P., Muster, T., Gill, H., Adhikaria, B., 2015. Enhanced efficiency fertilizers: a review of formulation and nutrient release patterns. J. Sci. Food Agric. 95, 1131−1142.

Trani, P.E., Terra, M.M., Tecchio, M.A., Teixeira, L.A.J., Hanasiro, J., 2013. Adubação orgânica de hortaliças e frutíferas. (In Dutch), Campinas: IAC.

Van Puijenbroek, P.J.T.M., Beusen, A.H.W., Bouwman, A.F., 2019. Global nitrogen and phosphorus in urban waste water based on the Shared Socio-economic pathways. J. Environ. Manag. 231, 446−456. https://doi.org/10.1016/j.jenvman.2018.10.048.

Wang, J., Vanga, S.K., Saxena, R., Orsat, V., Raghavan, V., 2018. Effect of climate change on the yield of cereal crops: a review. Climate 6 (2), 41.

Zaman, M., Zaman, S., Adhinarayanan, C., Nguyen, M.L., Nawaz, S., Dawar, K.M., 2013. Effects of urease and nitrification inhibitors on the efficient use of urea for pastoral systems. Soil. Sci. Plant Nut. 59 (4), 649−659.

Zerulla, W., Barth, T., Dressel, J., Erhardt, K., Von Locquenghien, K.H., Pasda, G., Radle, M., Wissemeier, A., 2001. 3,4-Dimethylpyrazole phosphate (DMPP) − a new nitrification inhibitor for agriculture and horticulture. Biol. Fertil. Soils 34, 79−84.

Zhang, F., Wang, Q., Hong, J., Chen, W., Qi, C., Ye, L., 2017. Life cycle assessment of diammonium-and monoammonium-phosphate fertilizer production in China. J. Clean. Prod. 141, 1087−1094.

# Nitrogen sinks in the agro-food system of Pakistan

3

**Sajjad Raza[1], Muhammad Arif Watto[2], Annie Irshad[3], Muhammad Nasim[4], Xiaoning Zhao[1]**

[1]*School of Geographical Sciences, Nanjing University of Information Science & Technology, Nanjing, Jiangsu, China;* [2]*University of Agriculture Faisalabad, Sub-Campus at Depalpur, Okara, Punjab, Pakistan;* [3]*College of Grassland Agriculture, Northwest A&F University, Yangling, Shaanxi, China;* [4]*Pesticide Quality Control Laboratory, Bahawalpur, Punjab, Pakistan*

## 3.1 Introduction

Plants primarily depend on inorganic nitrogen (N), which, after carbon and oxygen, is the most abundantly found element in plant dry matter, typically ranging between 10 and 30 g kg$^{-1}$ (McNeill and Unkovich, 2007). It is a key component of plant amino and nucleic acids and chlorophyll, and is usually acquired by plants in greater quantities from soil than any other element. Plant N is the basis for dietary protein for all animals and humans. World human population has been increased from 2.54 billion people in 1950 to nearly 8 billion people in 2020 (FAOSTAT, 2020). It is believed that about 50% increase in global food production is linked solely to N fertilization over the past five decades (Smil,2002; Brown, 1999) and nearly half of the world population is considered to be alive today because of N fertilizers (Zhang et al., 2013). It is being projected that N applications will further increase in order to meet food demands of the rapidly growing human population in the coming years.

Nitrogenous fertilizers have played a leading role in maximizing food production to meet the burgeoning demand of the rapidly increasing world population. During the past 6–8 decades, human interventions as a result of the industrial development have resulted in massive expansion of global N cycle mainly through substantial increase in reactive N input driven by industrial N fixation via Haber–Bosch process (Erisman et al., 2008). The magnitude of N fixation from anthropogenic sources has doubled relative to natural sources and greatly expanded the global N cycle over the last century (Fowler et al., 2013). Globally, around 103.7 million tons of N fertilizers are added to soils every year (IFASTAT, 2020). The rate of increase in N fertilization is higher in developing countries than in developed countries (Fig. 3.1). Pakistan alone accounts for approximately 3% of the global N fertilizers consumption (IFASTAT, 2020) which has increased by 54-fold since 1961 to 2017—from 62.1 Gg yr$^{-1}$ in 1961 to 3439 Gg yr$^{-1}$ in 2017 (Fig. 3.2). This makes an increase

Nitrogen Assessment. https://doi.org/10.1016/B978-0-12-824417-3.00003-4

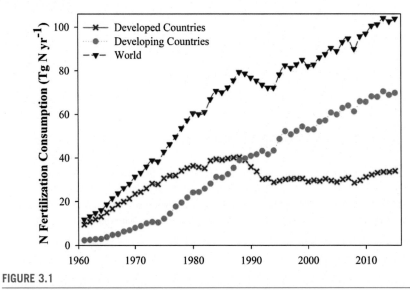

**FIGURE 3.1**

Global trends in nitrogen fertilizer consumption (Tg N yr$^{-1}$) in developed and developing countries (IFASTAT, 2020).

from 4.9 kg ha$^{-1}$ in 1960s to more than 150 kg ha$^{-1}$ in 2010 (Raza et al., 2018). In Pakistan, the share of N fertilizers is about 80% of the total nutrient consumption by the crop sector.

Nitrogen is a dynamic nutrient that exists in many forms and can move through different systems including soil, plant, water, and atmosphere. Nitrogen movement within these systems is regulated by various factors such as sources and sinks of N and drivers of N use. Nitrogen sources are the ways through which N is added to a system, whereas N sinks refer to the N flow to another part of its natural cycle where it mainly accumulates; however, it may flow to another part. A balance between N source and sink keeps N system at equilibrium. However, it becomes challenging and problematic when there is a continual loss of reactive N from one system and excessive accumulation in another system. Nitrogen-related environmental concerns have become prominent in the past few decades mainly because of the disequilibrium resulting from poor conversion of fertilizer N sources into plant N sinks. Such imbalances are bringing a cascade of environmental changes which are negatively affecting both living beings and ecosystems (Kanter et al., 2020).

Major N sources include synthetic fertilizer, atmospheric deposition, manures, biological N fixation, and crop residue recycling (More detail in Chapter 2). Main N sinks include (i) N uptake by plants; (ii) accumulation in soil profile; (iii) immobilization in microbial biomass or soil organic matter; (iv) losses to the environment in the form of various gases; (v) leaching into groundwater or disposal into rivers or oceans and; (vi) consumption by humans and livestock (Fig. 3.3). Knowing changes in N sources and sinks is important to (i) maintain N balance; (ii) know the progress

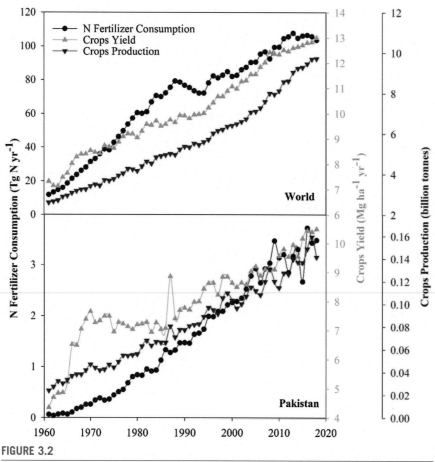

**FIGURE 3.2**

Comparison of historical changes in crops yield, production with N fertilization during 1961–2020 between the world and Pakistan.

of N transformations; (iii) predict N losses; and (iv) quantify N accumulation in different reservoirs (soil, air, and water) that are ultimately needed to reduce pollution and save ecosystem.

Rising N levels in sinks (like air and water resources) are polluting the environment. Gaseous N losses are accelerating climate change, global warming, and ozone layer depletion (Erisman et al., 2011). Ammonia ($NH_3$) released from soils through volatilization plays a key role in smog and formation of particulate matter which impairs air quality and affects human health (Wang et al., 2015). Events of nitrate contamination of groundwater resources, eutrophication, and soil acidification have accelerated and are directly affecting soil and human health, crop productivity, and degradation of ecosystem services (Galloway et al., 2003; Raza et al., 2020). Therefore, in order to devise policies to achieve environmentally sustainable food

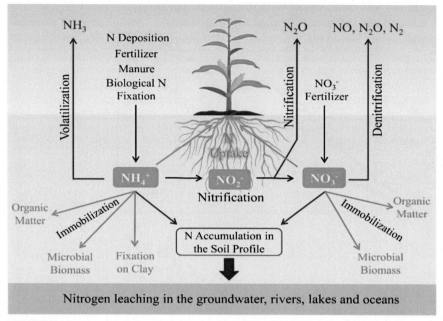

**FIGURE 3.3**

Nitrogen transformations showing N sinks and sources in an agricultural system. *Blue-colored text and arrows* shows N sources. *Red- and green-colored arrows* represent N sinks. *Red-colored text and arrows* represent harmful N sinks, while *green-colored arrows and text* represent useful N sinks. *Light blue—colored box* at the bottom represents N sinks in water bodies.

production worldwide, it is very important to understand N movement through different sources and sinks in agricultural systems.

This chapter focuses on the major N sinks and how these are cycled through different ecosystems in Pakistan. We tend to address the following questions: (i) how N sinks have changed over the past six decades in the agro-food system of Pakistan? (ii) How much of the N consumed by plants (N use efficiency (NUE)) is then used by humans and livestock? (iii) How much of the N applied to crops eventually ends up being lost to the environment?

## 3.2 Major N sinks in the agro-food systems

Nitrogen being mobile moves quickly through soil—plant—water—environment systems. During this movement, variable (high or low) quantities of N may accumulate in short- or long-term storage pools known as N sinks. Knowing the movement of N within different sinks is important to evaluate N balance, ecosystem sustainability, and environmental quality. Nitrogen accumulation and its movement within different sinks is explained in below subsections.

### 3.2.1 Plants N uptake

All plants require N for their growth and development. Globally, crop yield and fertilizer N consumption have increased linearly during the past six decades (Zhang et al., 2021) and the global crop production has increased from 2.6 billion tons $yr^{-1}$ in 1961 to 9.7 billion tons $yr^{-1}$ in 2018 with an increase in crop yields from 7.3 to 13.0 Mg $ha^{-1}$ $yr^{-1}$ over the same period (FAOSTAT, 2020). Wood et al. (2004) projected that 50%−70% more cereal grain would be required by 2050 to feed 9.3 billion people with a projected 50%−70% increase in N fertilizers. Pakistan has also recorded a substantial increase in crop production and yield over the past six decades. In 1961, total crop production was only 24 million tons which has increased consistently with time and reached 159 million tons in 2017 (FAOSTAT, 2020). Similarly, average yield of all crops has increased from 4.73 to 10.5 tons $ha^{-1}$ over the same period. Globally, increase in crop yield and production is attributed to substantial increase in N fertilization as well as in Pakistan. Since majority of the applied N fertilizers are consumed by plants, therefore, plants act as a major sink for N.

Nitrogen consumption by plants is termed as NUE which is a mass balance principle and is defined as N recovered in the harvested crop divided by the N applied to that crop. NUE generally indicates ecosystem productivity and sustainability of agricultural systems and also it indicates the surplus N as potential threat to the environment. In terms of NUE, a sustainable agriculture system should maintain high proportion of N consumption by crops and minimum level of N loss to the environment. It is suggested that highly productive agro-ecosystems should exhibit N uptake of 60%−70% of the total N input (Conant et al., 2013). However, the recovery of N fertilizers by plants generally remains far less as compared to the desired levels. Globally, it is estimated that out of the total N applied, about 25%−45% is actually consumed by crops and the rest 55%−75% goes as surplus (Lassaletta et al., 2014). Nitrogen uptake efficiency in developing countries like China, India, and Pakistan is very low (less than 35%) (Raza et al., 2018; Zhang et al., 2015). During 1961−65, about 58% of the N applied to cropland in Pakistan was recovered by the harvested plants, whereas 42% was reported to be as surplus N. The NUE has consistently decreased to 23% during 2009−13. Consequently, the annual N surplus has gone up to 77% (3581 Gg N) in Pakistan (Raza et al., 2018). Surplus N moves to other sinks such as N accumulation in soil profile and its losses in the atmosphere and in water resources which are discussed subsequently in this chapter.

Nitrogen uptake (sink) by crops in Pakistan showed an increasing trend during the past six decades increasing from 237 Gg N $yr^{-1}$ during 1961−65 to more than 1055 Gg N $yr^{-1}$ during 2013−18 (Table 3.1). Nitrogen uptake in different crop species in Pakistan varied widely with cereals being the largest N sink (about 60%)

**Table 3.1** Historical changes in N uptake (sink) in important crops in Pakistan from 1961 to 2018.

| Crops | Average crop N uptake (Gg N yr$^{-1}$) | | |
|---|---|---|---|
| | 1961–65 | 1980–85 | 2013–18 |
| Cereals | 130 | 188 | 655 |
| Oilseeds | 62 | 110 | 314 |
| Fiber crops | 0.4 | 0.8 | 2 |
| Sugar crops | 5 | 7 | 18 |
| Roots/tubers | 0.4 | 3 | 10 |
| Fruits | 2 | 4 | 9 |
| Vegetables | 2 | 3.5 | 10 |
| Leguminous crops | 32 | 32 | 29 |
| Other crops | 4 | 5 | 8 |
| Total | 237 | 353 | 1055 |

*Reproduced from FAOSTAT, 2020. Statistics Division (Rome: Food and Agriculture Organization of the United Nations). www.fao.org/faostat/en/#data; Raza, S., Zhou, J., Aziz, T., Afzal, M. R., Ahmed, M., Javaid, S., Chen, Z., 2018. Piling up reactive nitrogen and declining nitrogen use efficiency in Pakistan: a challenge not challenged (1961–2013). Environ. Res. Lett. 13 (3), 034012.*

followed by fiber crops (20%), sugar crops (8%), fruits (3.5%), and vegetables (2%) (Raza et al., 2018). The contribution of cereals as N sink has substantially increased during the past six decades from 130 to 655 Gg N yr$^{-1}$ during 1961–2013. The contribution of leguminous crops in N sink was 14% which has decreased over the years because of a gradual decrease in area under cultivation and production of leguminous crops. The data show that N uptake by leguminous crops contributed less than 3% of the total N consumption by crops in Pakistan during 2009–13 (Table 3.1).

Nitrogen sink by crop plants depend on several factors such as (1) nutrients supply capacity of soil; (2) plant genetic makeup; (3) root depth and architecture; (4) plant growth and yield potential; (5) N assimilation capacity of roots and its utilization by plants; and (6) duration of a crop cycle. Improving plants' ability to utilize N is a key component in enhancing production and maintaining environmental sustainability.

## 3.2.2 Livestock sector

Nitrogen also plays an important role in animal production. Demand for animal feed produced from different crops and by-products of various food industries has rapidly increased in the past century. Recent estimates suggest that 4.7–7.0 billion tons dry biomass is being consumed by livestock each year which is equivalent to nearly 60% of the global plant biomass (Herrero et al., 2008; Krausmann et al., 2008; Wirsenius et al., 2010). About 34% of the world cereals are used as animal feed and about 30% of global arable land is used for producing livestock feed, probably also involving a similar amount of fertilizer use (FAO, 2013; Steinfeld et al., 2006). Thus, majority of the N stored in plants is eventually consumed by livestock and thus making it an important N sink. In addition, growing livestock population will require allocation of more land, crops, and other factors of production, thereby driving source and sink of N cycle in many ways.

The volume of N sink in livestock sector has increased worldwide over the past few decades because of the substantial increases in the number of live animals (FAO-STAT, 2020) which is well correlated with food demand, and changes in dietary patterns. Based on historical data assessments, it is being reported that improvements in living standards have resulted in changed dietary patterns of human being with higher demands for animal protein as compared to the plant-based proteins (Lassaletta et al., 2016). These impending changes will cause further expansion in N sink in livestock systems. Livestock sector showed an enormous growth during the past six - decades worldwide as well as in Pakistan where a considerable increase in poultry and livestock population was observed, particularly for chicken, goat, and buffalo (Table 3.2). To maintain supply of feed to increasing livestock population, the portion of grain crops derived N has also been increased, which makes livestock as an important N sink in Pakistan. Animal feed production in Pakistan is not sufficient enough to feed the increasing livestock population and, hence, the deficit is fulfilled through import of feed products of maize, barley, and soybean derivatives (Raza et al., 2018). Among livestock products, majority of the N sink was found in milk and milk products which increased from 43 Gg N yr$^{-1}$ in 1961 to 257 Gg N yr$^{-1}$ in 2013 (Table 3.3). Meat of ruminants and poultry and hides/skins were the other major components of N sink by livestock.

**Table 3.2** Changes in live animal stocks in Pakistan from 1960 to 2018.

| Animal species | Live animals stock (millions) | | | | | | | Relative increase (%) (1961−2018) |
|---|---|---|---|---|---|---|---|---|
| | 1961 | 1970 | 1980 | 1990 | 2000 | 2010 | 2018 | |
| Buffaloes | 6.7 | 9.3 | 11.5 | 17.4 | 22.7 | 29.4 | 38.8 | 480 |
| Cattle | 14.2 | 14.6 | 15.0 | 17.7 | 22.0 | 34.3 | 46.1 | 225 |
| Chickens | 11.5 | 16.8 | 45.2 | 79.0 | 150 | 321 | 524 | 4458 |
| Ducks | 0.3 | 0.4 | 1.3 | 2.8 | 3.5 | 3.7 | 3.8 | 1180 |
| Goats | 8.8 | 13.2 | 25.0 | 35.4 | 47.4 | 59.9 | 74.1 | 742 |
| Sheep | 10.2 | 13.1 | 21.4 | 25.7 | 24.1 | 27.8 | 30.5 | 198 |
| Horses | 0.4 | 0.4 | 0.5 | 0.4 | 0.3 | 0.4 | 0.4 | 5.4 |
| Mules | 0.0 | 0.0 | 0.1 | 0.1 | 0.2 | 0.2 | 0.2 | 645 |
| Asses | 1.0 | 1.7 | 2.4 | 3.4 | 3.8 | 4.6 | 5.3 | 460 |
| Camels | 0.6 | 0.7 | 0.8 | 1.0 | 0.8 | 1.0 | 1.1 | 79 |
| Beehives | 0.1 | 0.1 | 0.1 | 0.1 | 0.1 | 0.4 | 0.4 | 649 |

*Data from FAOSTAT, 2020. Statistics Division (Rome: Food and Agriculture Organization of the United Nations). www.fao.org/faostat/en/#data.*

**Table 3.3** Historical changes in N (sink) in different livestock products in Pakistan (Raza et al., 2018).

| | Livestock N uptake (Gg N yr$^{-1}$) | | |
|---|---|---|---|
| **Livestock** | **1961–65** | **1980–85** | **2009–13** |
| Meat ruminants | 9 | 18 | 48 |
| Meat poultry | 0.2 | 6 | 15 |
| Milk and products | 43 | 89 | 257 |
| Hides/skins | 12 | 29 | 61 |
| Others | 0 | 0.3 | 1 |
| Total | 65 | 142.3 | 381 |

Conversion of plant protein into animal protein in livestock production system is highly inefficient and only 5%–45% of N in plant protein is converted into animal protein depending on animal species and their ration and management. The other 55%–95% is excreted via urine and feces, which can be used as a nutrient source for plants production (Oenema and Tamminga, 2005). Production of 1 kg of chicken, pork, and beef require about 2, 4, and 8 kg of cereals, respectively (Oenema and Tamminga, 2005). In this way, livestock sector also acts as a major N source and more detail about this is discussed in Chapter 2.

## 3.3 Humans food as N sink

Nitrogen is an important structural component of proteins and it plays a significant role in a number of vital physiological and biochemical processes sustaining human life. Nitrogen fertilizers have played a tremendous role in maximizing world food production and are a mainstay of half of the world's population against hunger (Zhang et al., 2013). The recommended protein intake for adults is 0.8–0.9 g kg$^{-1}$ of body weight d$^{-1}$ (Phillips, 2012) and recommended annual N intake is 2.5–3.5 kg N capita$^{-1}$ yr$^{-1}$ (WHO, 2007). Recommended intakes are expressed in terms of ideal protein containing adequate amounts of all essential amino acids and are easily digestible.

Crops and livestock are the main protein sources for human intake worldwide. Pakistan is the fifth-most populous country in the world and its population has increased from 37.5 million in 1947 to 212.7 million in 2020 (FAOSTAT, 2020). Increasing population is adding more mouths for N intake and adding pressure to produce more protein from crop and livestock sources. Nitrogen sink by humans has consistently increased in Pakistan since the past six decades and is expected

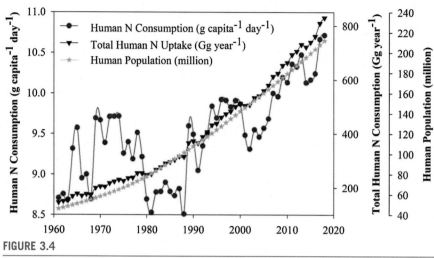

**FIGURE 3.4**

Changes in human N intake per person (g capita$^{-1}$ d$^{-1}$) and total human N uptake in Pakistan.

*Based on data from FAOSTAT, 2020. Statistics Division (Rome: Food and Agriculture Organization of the United Nations). www.fao.org/faostat/en/#data.*

to further increase in the future (Fig. 3.4). Total human N consumption in Pakistan was 146 Gg N yr$^{-1}$ in 1961 which slowly increased to 248 Gg N yr$^{-1}$ till 1980. After that, an abrupt increase in N consumption was observed in the next 4 decades, reaching 830 Gg N yr$^{-1}$ in 2018. A major reason behind this substantial increase in N consumption was rapidly increasing human population. The N consumption at individual level was also increased from 8.71 to 10.72 g capita$^{-1}$ d$^{-1}$ over the same period.

Pakistani people fulfill 98% of their N requirements by consuming crops and livestock products (Fig. 3.5). Major contribution for protein intake comes from crops. However, crops-derived N consumption by humans has decreased from >70% during 1961%−74% to 58% in 2017 (Fig. 3.5). On the other hand, the contribution of livestock in human N consumption was about 24%−27% during 1964−74 which has gone beyond 41%, whereas a further increase is expected in the future (Fig. 3.5).

Livestock and humans also act as a source of N by its excretion in the environment. The increasing livestock and human populations and their increasing N intakes are also contributing in higher N excretions and thus more losses to the environment. More detail about this is discussed in Chapter 2.

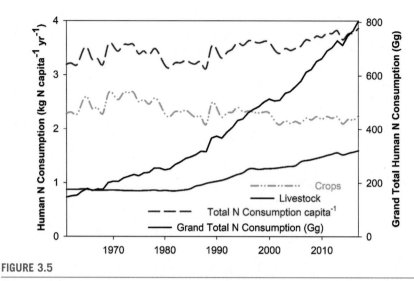

**FIGURE 3.5**

Changes in human N intake in Pakistan from different food sources during 1961–2017.

*Data from FAOSTAT, 2020. Statistics Division (Rome: Food and Agriculture Organization of the United Nations).*

*www.fao.org/faostat/en/#data.*

## 3.4 Atmosphere as N sink

The concentration of di-nitrogen ($N_2$) in the atmosphere is about 78% which is the major reservoir of N. The di-nitrogen is inert and thus not responsible for any environmental damages. However, fixation of atmospheric N, both natural and anthropogenic, contributes toward production of reactive N in various forms. With industrial revolution and intensive agricultural practices, the gaseous N losses to the atmosphere have grown rapidly during the past six decades, degrading air quality and causing other environmental and health issues (Xuejun and Fusuo, 2011; McLauchlan et al., 2013). Nitrogen leakages from the use of fertilizer is the major source of gaseous N emissions to the atmosphere. Gaseous N losses generally occur in the form of ammonia ($NH_3$) volatilization, nitrous oxide ($N_2O$), nitric oxide (NO), and nitrogen dioxide ($NO_2$) collectively termed as NOx, and dinitrogen ($N_2$). Gaseous N emissions reduce ecosystem productivity. Gaseous N losses as $NH_3$ from soils and NOx form other combustion-related emissions contribute about 100 Tg N $yr^{-1}$ of N losses worldwide (Fowler et al., 2013). These gaseous emissions are processed within the atmosphere and get transported between countries generating secondary pollutants, such as ozone and other photochemical oxidants and aerosols, threatening air quality.

Soils are an important source of gaseous N emissions mainly derived from the huge amount of manure (24.5 Tg N $yr^{-1}$) and synthetic N fertilizer (104 Tg N $yr^{-1}$) added to soils (Zhang et al., 2017; IFASTAT, 2020). Recent estimates have shown a significant increase in gaseous N emissions from 82[1] to 1216 Gg N $yr^{-1}$ during the past six decades in Pakistan (Raza et al., 2018). These emissions on the national scale can be correlated with total use of N fertilizers on arable lands. Air quality has been deteriorated in major cities particularly in Lahore, Karachi, and Faisalabad where AQI remains significantly high. During winter, smog problem is common in many major cities, particularly in the central and southern Punjab and Sindh. The worsening air quality is particularly due to increasing concentrations of $NH_3$, $N_2O$, and NO in Pakistan (Iqbal and Goheer 2008; Khan et al., 2011).

### 3.4.1 Ammonia volatilization

Ammonia volatilization is a chemical process that occurs at the soil surface when ammonium ion ($NH_4^+$) is converted into ammonia gas and is released into the atmosphere. Agriculture is the largest source of $NH_3$ in the atmosphere and global agricultural $NH_3$ emissions reported to be increased by 90% between 1970 and 2005 (Sailesh et al., 2013). $NH_3$ volatilization from soils in response to N fertilization is 12 Tg N $yr^{-1}$ worldwide (Riddick et al., 2016). Overall, more than 40% of the applied N is reported to be lost as $NH_3$ under certain environmental and edaphic conditions (Singh et al., 2013). On average, between 10% and 14% N is lost via volatilization from synthetic fertilizers (Bouwman et al., 2002; Klein et al., 2006) and losses may reach up to 50% of N applied to rain-fed agricultural systems particularly on alkaline soils and as much as 80% in flooded agricultural systems (Freney, 1997). Surface application of $NH_4^+$ containing fertilizers or farm yard manure is a predominant cause of $NH_3$ volatilization from intensive agricultural systems (Pan et al., 2016; Van der Stelt et al., 2007). Broadcasting N fertilizers without incorporation increases the susceptibility to $NH_3$ loss (Soares et al., 2012). In general, higher pH, warmer temperatures, and greater soil moisture content increase the potential for volatilization (Sanz-Cobena et al., 2011). After reaction with atmospheric acidic substances, $NH_3$ produces particulate matter which contributes to haze pollution causing serious impacts on human health and also causes reduced visibility affecting the transport sector (Gong et al., 2013).

Application of N in the form of animal manure and excretion of urine and feces by grazing animals is a source of soil $NH_4^+$ which can rapidly be converted into $NH_3$. Global $NH_3$ emissions from animal manure are about 21 Tg N $yr^{-1}$ and out of that 44% come from animal houses and storage systems, 27% from grazing animals, and 30% from spreading of animal manure (Beusen et al., 2008). $NH_3$ volatilization losses from animal waste occur relatively rapidly and are influenced by a range of environmental and management factors (Bolan et al., 2004; Oenema et al., 2001).

Pakistan has a hot and dry climate with a maximum temperature in summer about 50 °C in central and southern Punjab and Sindh. Furthermore, majority of soils in Pakistan are calcareous in nature with pH generally above or around 8.0 (Maqsood et al., 2016). Such soils and climatic conditions are favorable for ammonia volatilization. Ammonia volatilization losses have considerably increased during the past six decades in the country (Fig. 3.6). Out of the total gaseous N emissions in Pakistan, ammonia volatilization constitutes a major share exceeding 83% during

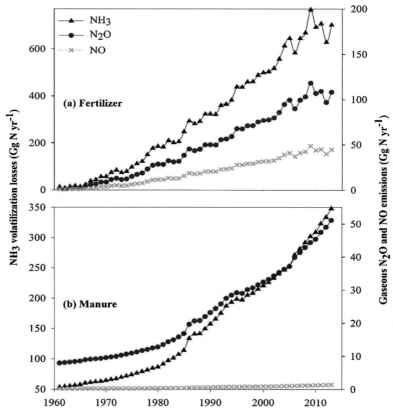

**FIGURE 3.6**

Gaseous emission of $N_2O$, NO, and $NH_3$ from (A) synthetic N fertilizers and (B) manure applied to soils in Pakistan during 1961–2013.

*Data from FAOSTAT, 2020. Statistics Division (Rome: Food and Agriculture Organization of the United Nations). www.fao.org/faostat/en/#data and calculated from Raza, S., Zhou, J., Aziz, T., Afzal, M. R., Ahmed, M., Javaid, S., Chen, Z., 2018. Piling up reactive nitrogen and declining nitrogen use efficiency in Pakistan: a challenge not challenged (1961–2013). Environ. Res. Lett. 13 (3), 034012.*

the past six decades (Raza et al., 2018). During 1961–65, the average total ammonia volatilization losses were only 70 Gg N $yr^{-1}$ which went beyond 1023 Gg N $yr^{-1}$ during 2009–13 (Fig. 3.6). This increase has been specifically faster after 1980, when the era of intensive N fertilization started in Pakistan. The share of N fertilizer in total ammonia volatilization was only 20% during 1961–65 which reached 68% during 2009–13 (Fig. 3.6).

### 3.4.2 Nitrogen oxides

Emissions of N oxides in the form of $N_2O$, NO, and $NO_2$ are driven by nitrification and denitrification processes. Nitrification is the process that converts ammonium

($NH_4^+$) into nitrite ($NO_2$) and then into nitrate ($NO_3$) (Norton, 2008). Mostly, nitrification occurs by ammonia-oxidizers and nitrite-oxidizers which are ubiquitous in aerobic environments (Di et al., 2009). Nitrification is a precursor of majority of N losses including $N_2O$ emissions and reduces N availability to plants by up to 50% (Beeckman et al., 2018). Microbial nitrification and denitrification in managed and natural soils contribute approximately 70% of global $N_2O$ emissions (Syakila and Kroeze, 2011; Braker and Conrad, 2011). The concentration of atmospheric $N_2O$ has increased to 331 ppb in 2018 with the fastest growth observed in the past five decades (Tian et al., 2020). $N_2O$ emissions from N fertilization and manure management ($4.3-5.8$ Tg $N_2O-N$ $yr^{-1}$) and emissions from natural soils ($6-7$ Tg $N_2O-N$ $yr^{-1}$) contribute 56%$-$70% of all global $N_2O$ sources (Syakila and Kroeze, 2011).

The soil and climate conditions in Pakistan are generally suitable for nitrification because of high temperature and pH (Maqsood et al., 2016). Therefore, high nitrification rates generally accelerate losses of $N_2O$ and NO because it accelerates the denitrification. Loss of $N_2O$ and NO emissions in Pakistan from synthetic fertilizer applications has been increased from 10 to 1 Gg N $yr^{-1}$ during 1961$-$65 to 147 and 46 Gg N $yr^{-1}$ during 2009$-$13, respectively (Fig. 3.6) (Raza et al., 2018). The share of fertilizer in combined $N_2O$ and NO emissions has increased from 27% during 1961%$-$65% to 77% during 2009$-$13 (Fig. 3.6). Moreover, the share of $N_2O$ and NO emissions in total gaseous N losses was 14% during 1961$-$65 and increased to 16% during 2009$-$13. Overall, the total gaseous N emissions including $NH_3$, $N_2O$, and NO were 20% (82 Gg N) of the total N input applied to cropland during 1961$-$65 which reached 26% (1216 Gg N) during 2009$-$13 (Raza et al., 2018). It is important to control gaseous N losses because $N_2O$ is one of the most important greenhouse gases contributing to climate change. $N_2O$ is also a major source of ozone-depleting nitric oxide (NO) and nitrogen dioxide ($NO_2$) in the stratosphere.

Unlike nitrification, denitrification is an anaerobic process which usually occurs in soils and sediments and anoxic zones in lakes and oceans. Denitrification converts nitrate into $NO_2$, NO, $N_2O$, and $N_2$ gas (Šimek et al., 2002). Terrestrial soil denitrification is the major global N sink because of its substantial role in closing N cycle through the conversion of about 30%$-$60% of Nr back to atmospheric as $N_2$ (Seitzinger et al., 2006, Stocker, 2014). Contributions of gaseous N losses driven by denitrification are generally low in Pakistan mainly because the most parts of the country fall in dryland agroecosystem with high temperature and soils are calcareous.

## 3.5 Soil as N sink

Generally, soils are deficient in terms of available N and that is mainly due of low organic matter contents. Therefore, application of N fertilizers and/or manures is compulsory in order to increase crop yields. Once nitrogen is added into the soil, it becomes available for plant uptake; however, plant N uptake from soil in many cases does not exceed 50% of total N applied. The surplus N can cause accumulation of significant amounts of N in soils (Ju et al., 2004; Raza et al., 2019). Reduced NUE and high nitrification potential of soils facilitate $NO_3^-$ accumulation in agricultural soils (Zhang et al., 2013).

Nitrogen accumulation in the soil profile can be in three major N pools: (1) dissolved $NO_3^-$ and $NH_4^+$ in the vadose zone; (2) in groundwater aquifers; and (3) organic N storage within the soil profile (Van Meter et al., 2016). Indeed, the largest pool of N in most terrestrial ecosystems is soil organic N (SON) (Galloway et al., 2003). It is estimated that terrestrial N sequestration may occur at a global scale in the order of $20-100$ Tg N yr$^{-1}$ (Fowler et al., 2013; Zaehle, 2013). Yan et al. (2014) found that the average soil N content of Chinese croplands increased by 5.1% between 1979 and 82 to 2007−08 while mass balance and modeling studies in Canada (Clair et al., 2014), Europe (Leip et al., 2011), and the United States (SAB, 2011) report an annual accumulation of N within agricultural soils on the order of 15%−20% of total N (TN) inputs.

An increase in the accumulation of N in soils is a direct result of increased N input from all sources (inorganic and manure N). N input from all sources increased from 0.4 Tg N yr$^{-1}$ (1961−65) to 4.6 Tg yr$^{-1}$ (2009−13) during the past six decades in Pakistan. Inefficient N utilization and poor crop management practices result in very low N uptake by crops and higher amounts of surplus N in soils. Crops N utilization efficiency in Pakistan decreased from 58% during 1961−65 to 23% during 2009−13 and the concurrent N surplus increased from 0.17[1] to 3.6 Tg N yr$^{-1}$ during the same time period.

Based on mass-balance approach, after minimizing N uptake by crops, and gaseous N losses, it is estimated that around 89 Gg N yr$^{-1}$ was available for accumulation in soils of Pakistan during 1961−65 which has considerably increased during the past 6 decades. It is estimated that during 2009−13 around 2.3 Tg N yr$^{-1}$ was available for accumulation in soils of Pakistan (Table 3.4). This is an amount of N available for accumulation which can have different fates: (i) a part of this will eventually get accumulated in soils; (ii) move downward and contaminate the groundwater aquifers; (iii) move with water in the form of riverine export and become part of rivers and lakes; and (iv) can be lost to the atmosphere.

The climate, soil type, and N fertilization scenarios in Pakistan facilitate N accumulation and leaching into the soil profile. Nitrification process is very fast in majority of the soils in Pakistan; hence, most of the applied urea or other $NH_4^+$ fertilizers are quickly nitrified releasing nitrates. The net charge in soils is negative which repel negatively charged nitrate and therefore facilitate its movement downward the soil profile at greater depths. Although precipitation is low, but majority of the rainfall (>50%) is usually received in 4 months from June to September during the monsoon season. Occasionally, heavy rainfall in this season not only causes N losses by surface runoff but also transports surface $NO_3^-$ deep into the soil profile.

**Table 3.4** Mass-balance estimation of N available for accumulation in soils of Pakistan during 1961–2013.

| Year | Total N input (Gg yr$^{-1}$) | Crops N uptake (Gg yr$^{-1}$) | Surplus N (Gg yr$^{-1}$) | Total gaseous N losses (Gg yr$^{-1}$) | N Available for accumulation in soil (Gg yr$^{-1}$) |
|---|---|---|---|---|---|
| 1961 | 385.9 | 216.0 | 169.9 | 78.7 | 91.2 |
| 1962 | 371.5 | 227.5 | 143.9 | 73.9 | 70.0 |
| 1963 | 413.9 | 244.3 | 169.6 | 82.5 | 87.1 |
| 1964 | 435.9 | 241.4 | 194.5 | 88.3 | 106.2 |
| 1965 | 432.4 | 256.2 | 176.2 | 85.3 | 90.9 |
| 2009 | 4874.0 | 1102.1 | 3771.9 | 1281.4 | 2490.6 |
| 2010 | 4551.3 | 1003.5 | 3547.8 | 1199.7 | 2348.1 |
| 2011 | 4672.2 | 1081.7 | 3590.5 | 1233.8 | 2356.7 |
| 2012 | 4344.7 | 1014.4 | 3330.4 | 1150.6 | 2179.8 |
| 2013 | 4735.4 | 1072.5 | 3662.9 | 1255.6 | 2407.4 |

*Calculated from Raza, S., Zhou, J., Aziz, T., Afzal, M. R., Ahmed, M., Javaid, S., Chen, Z., 2018. Piling up reactive nitrogen and declining nitrogen use efficiency in Pakistan: a challenge not challenged (1961–2013). Environ. Res. Lett. 13 (3), 034012.*

## 3.6 Water resources as N sink

Besides accumulating in soils, a large proportion of the anthropogenically mobilized N eventually either enters into groundwater aquifers or is transported through freshwater toward coastal marine systems. N leaching into groundwater and mixing into freshwater results in numerous negative impacts on human health as well as on the environment including surface and groundwater pollution, loss of habitat and biodiversity, increase in frequency and severity of harmful algal blooms, eutrophication, and hypoxia affecting aquatic life (Erisman et al., 2013).

Numerous studies have reported nitrate contamination in groundwater resources throughout Pakistan with synthetic N fertilizer applications being highlighted as the main cause. During 2007–08, the results of a study with 747 samples (from surface and groundwater) taken from the entire country reveal that 19% of the samples contained nitrate concentrations (23% in Balochistan and Punjab provinces) beyond the safe limits (Tahir and Rasheed, 2008). However, the number went up to 23% in 2017 (Podgorski et al., 2017) (Fig. 3.7) with an additional 12% samples on the margin of nitrate contamination (7–10 mg L$^{-1}$) which can further make the total number go beyond 35% in near future. Various other studies have also reported nitrate contamination in different cities of the country like Kasur (Farooqi et al., 2007), Islamabad and Rawalpindi (Kazmi and Khan, 2005), Quetta and Faisalabad (PCRWR, 2005), and Lahore (Naeem et al., 2007).

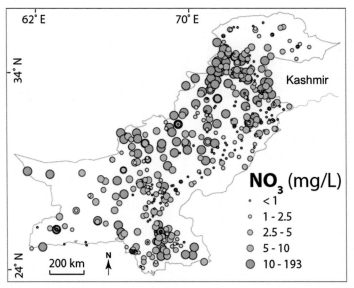

**FIGURE 3.7**

Inverse-distance-weighting grids of $NO_3^-$ in groundwater samples collected throughout Pakistan (n = 458).

*Adapted and reproduced from Podgorski, J.E., Eqani, S., Khanam, T., Ullah, R., Shen, H., Berg, M., 2017.*
*Extensive arsenic contamination in high-pH unconfined aquifers in the Indus Valley. Sci. Adv. 3 (8), 1–10.*

Surplus N via run off flows from soils and enters in streams, lakes, rivers, and eventually in the oceans. The increasing N levels in the water bodies are causing several challenges such as poor drinking water quality and eutrophication (Huang et al., 2017). Nitrogen export to riverine environment is in the dissolved and particulate forms. The dissolved forms include nitrate, nitrite, and ammonium, collectively called as dissolved inorganic nitrogen (DIN). Dissolved organic nitrogen (DON) may also have several forms and distribution such as components of dissolved organic matter. Global flux of dissolved N from all rivers is estimated to be around 4.45 Tg yr$^{-1}$ as DIN and 10 Tg yr$^{-1}$ as DON (Green et al., 2004).

The Indus River is a transboundary river that flows through four countries: China, Afghanistan, Pakistan, and India with >60% of its drainage area in Pakistan. The Indus River Basin faces severe water quality degradation because of nutrient enrichment from human activities. It has been estimated that export of dissolved N by the Indus River will increase by a factor of 1.6–2 between 2010 and 2050 (Wang et al., 2019). Agriculture and human waste are important drivers of N inputs to the basin (Fig. 3.8).

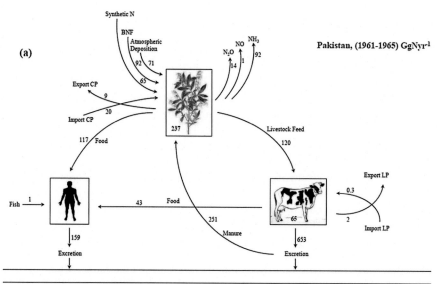

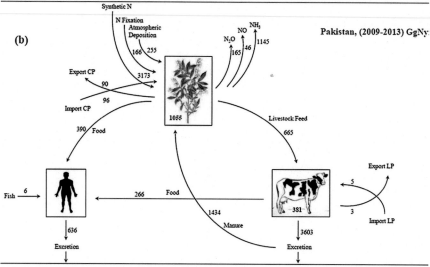

**FIGURE 3.8**

Major changes (Gg N yr$^{-1}$) in N sinks in Pakistan's agro-food system. The subfigures
(a) and (b) represent average N sinks during 1961—1965 and 2009—2013, respectively.

*Adapted from Raza, S., Zhou, J., Aziz, T., Afzal, M. R., Ahmed, M., Javaid, S., Chen, Z., 2018. Piling up reactive
nitrogen and declining nitrogen use efficiency in Pakistan: a challenge not challenged (1961—2013). Environ.
Res. Lett. 13 (3), 034012.*

## 3.7 Conclusion

Over the last few decades, nitrogen has played a substantial role in the global food and feed production. However, its excessive and imbalanced use is disturbing the natural nitrogen cycle which is threatening the environmental sustainability. It has been witnessed that with the passage of time, crop output to N input ratio is gradually declining globally as well as in Pakistan. As a result of the decreasing NUE, a huge amount of surplus N is being produced which is added to different N sinks. Nitrogen applications to agro-food systems of Pakistan have massively increased after the green revolution of 1960s. Plants N uptake is the major N sink but its volume has considerably decreased from 54% to 23% and is leading to higher Nr concentrations into the atmosphere, soils, and water resources which can have serious repercussions to environmental sustainability in the country. Although efforts on decreasing N-related environmental problems has also increased over the time, inefficient use of N and its losses from agricultural and livestock systems, however, still remains a big challenge. Pakistan can emerge as a hotspot of N pollution if N losses continue to progress at the current scale. Given this challenge, it is imperative to take proactive measures suing a holistic approach encompassing advanced agronomic practices and genetic improvements to reduce N losses and improve NUE.

Improving NUE at farm level requires a strong nexus among farmers, researchers, politicians, and extension efforts are needed to improve crop cultivation practices. There is need to promote balanced fertilization, and more focus on the adoption of "4R" principle approach for fertilizer use (right source, right amount, right time, and right placement). Existing cultivars should be replaced with highly efficient N utilizing cultivars. Soil testing should be increased and fertilization should be recommended based on the existing nutrients status of soil. Government should encourage fertilizer companies to initiate research and development activities into optimal N use (and balanced compound NPK), and to develop affordable enhanced-efficiency fertilizers such as those containing nitrification and urease inhibitors.

## References

Beeckman, F., Motte, H., Beeckman, T., 2018. Nitrification in agricultural soils: impact, actors and mitigation. Curr. Opin. Biotechnol. 50, 166–173.

Beusen, A.H.W., Bouwman, A.F., Heuberger, P.S.C., Drecht, G.V., Van Der Hoek, K.W., 2008. Bottom-up uncertainty estimates of global ammonia emissions from global agricultural production systems. Atmos. Environ. 42 (24), 6067–6077.

Bolan, N.S., Saggar, S., Luo, J., Bhandral, R., Singh, J., 2004. Gaseous emissions of nitrogen from grazed pastures: processes, measurements and modelling, environmental implications, and mitigation. Adv. Agron. 84, 37–120.

Bouwman, A.F., Boumans, L.J.M., Batjes, N.H., 2002. Estimation of global $NH_3$ volatilization loss from synthetic fertilizers and animal manure applied to arable lands and grasslands. Global Biogeochem. Cycles 16 (2), 8-1-8-14.

Braker, G., Conrad, R., 2011. Diversity, structure, and size of $N_2O$-producing microbial communities in soils—what matters for their functioning? Adv. Appl. Microbiol. 75, 33–70.

Brown, L.R., 1999. Feeding nine billion. In: Brown, L.R., Flavin, C., French, H. (Eds.), State of the World 1999. A World Watch Institute Report on Progress Toward a Sustainable Society. W. Norton and Company, NY, pp. 115—132.

Clair, T.A., Nathan, P., Shabtai, B., Adrian, L., Paul, A., Michael, D.M., Ian, D., David, N., Shannon, S., Craig, F.D., Jingyi, Y., 2014. Interactions between reactive nitrogen and the Canadian landscape: a budget approach: Canadian nitrogen budget. Global Biogeochem. Cycles 28 (11), 1343—1357.

Conant, R.T., Berdanier, A.B., Grace, P.R., 2013. Patterns and trends in nitrogen use and nitrogen recovery efficiency in world agriculture. Global Biogeochem. Cycles 27 (2), 558—566.

Di, H.J., Cameron, K.C., Shen, J.P., Winefield, C.S., O'Callaghan, M., Bowatte, S., He, J.Z., 2009. Nitrification driven by bacteria and not archaea in nitrogen-rich grassland soils. Nat. Geosci. 2 (9), 621—624.

Erisman, J.W., Sutton, M.A., Galloway, J., Klimont, Z., Winiwarter, W., 2008. How a century of ammonia synthesis changed the world? Nat. Geosci. 1 (10), 636—639.

Erisman, J.W., Galloway, J., Seitzinger, S., Bleeker, A., Butterbach-Bahl, K., 2011. Reactive nitrogen in the environment and its effect on climate change. Curr. Opin. Environ. Sustain. 3 (5), 281—290.

Erisman, J.W., Galloway, J.N., Seitzinger, S., Bleeker, A., Dise, N.B., Petrescu, A.R., Vries, W., 2013. Consequences of human modification of the global nitrogen cycle. Phil. Trans. Biol. Sci. 368 (1621), 20130116.

FAO, 2013. Food and Agriculture Organization of the United Nations. Food outlook. http://www.fao.org/docrep/018/al999e/al999e.pdf.

FAOSTAT, 2020. Statistics Division (Rome: Food and Agriculture Organization of the United Nations). www.fao.org/faostat/en/#data.

Farooqi, A., Masuda, H., Firdous, N., 2007. Toxic fluoride and arsenic contaminated groundwater in the Lahore and Kasur districts, Punjab, Pakistan and possible contaminant sources. Environ. Pollut. 145 (3), 839—849.

Fowler, D., Coyle, M., Skiba, U., Sutton, M.A., Cape, J.N., Reis, S., et al., 2013. The global nitrogen cycle in the twenty-first century. Phil. Trans. Biol. Sci. 368 (1621), 20130164.

Freney, J.R., 1997. Strategies to reduce gaseous emissions of nitrogen from irrigated agriculture. Nutrient Cycl. Agroecosyst. 48, 155—160.

Galloway, J.N., Aber, J.D., Erisman, J.W., Seitzinger, S.P., Howarth, R.W., Cowling, E.B., Cosby, B.J., 2003. The nitrogen cascade. Bioscience 53 (4), 341—356.

Gong, L.W., Lewicki, R., Griffin, R.J., Tittel, F.K., Lonsdale, C.R., Stevens, R.G., Pierce, J.R., Malloy, Q.G.J., Travis, S.A., Bobmanuel, L.M., Lefer, B.L., Flynn, J.H., 2013. Role of atmospheric ammonia in particulate matter formation in Houston during summertime. Atmos. Environ. 77, 893—900.

Green, P., Vorosmarty, C.P., Meybeck, M., Galloway, J., Boyer, E., 2004. Pre-industrial and contemporary N fluxes through rivers. Biogeochemistry 68, 7.

Herrero, M., Thornton, P.K., Kruska, R., Reid, R.S., 2008. Systems dynamics and the spatial distribution of methane emissions from African domestic ruminants to 2030. Agric. Ecosyst. Environ. 126 (1—2), 122—137.

Huang, J., Xu, C.C., Ridoutt, B.G., Wang, X.C., Ren, P.A., 2017. Nitrogen and phosphorus losses and eutrophication potential associated with fertilizer application to cropland in China. J. Clean. Prod. 159, 171—179.

IFASTAT, 2020. Nitrogen Statistics from IFADATA Statistics. http://ifadata.fertilizer.org/ucSearch.aspx.

Iqbal, M.M., Goheer, M.A., 2008. Greenhouse gas emissions from agro-ecosystems and their contribution to environmental change in the Indus Basin of Pakistan. Adv. Atmos. Sci. 25, 1043−1052.

Ju, X., Liu, X., Zhang, F., Roelcke, M., 2004. Nitrogen fertilization, soil nitrate accumulation, and policy recommendations in several agricultural regions of China. AMBIO: A J. Human Environ. 33 (6), 300−305.

Kanter, D.R., Bartolini, F., Kugelberg, S., Leip, A., Oenema, O., Uwizeye, A., 2020. Nitrogen pollution policy beyond the farm. Nat. Food 1 (1), 27−32.

Kazmi, S.S., Khan, S.A., 2005. Level of nitrate and nitrite contents in drinking water of selected samples received at AFPGMI Rawalpindi. Pak. J. Physiol. 1 (2), 28−31.

Khan, A.N., Ghauri, B.M., Jilani, R., Rahman, S., 2011. Climate Change: Emissions and Sinks of Greenhouse Gases in Pakistan Proceedings of the Symposium on Changing Environmental Pattern and its Impact with Special Focus on Pakistan (Lahore, Pakistan).

Klein, D.C., Novoa, R.S., Ogle, S., Smith, K., Rochette, P., Wirth, T., McConkey, B., Mosier, A., Rypdal, K., Walsh, M., 2006. $N_2O$ emissions from managed soils, and $CO_2$ emissions from lime and urea application. In: Eggleston, H.S., Buendia, L., Miwa, K., Ngara, T., Tanabe, K. (Eds.), IPCC Guidelines for National Greenhouse Gas Inventories. IGES, Hayama, Japan.

Krausmann, F., Erb, K.-H., Gingrich, S., Lauk, C., Haberl, H., 2008. Global patterns of socio-economic biomass flows in the year 2000: a comprehensive assessment of supply, consumption and constraints. Ecol. Econ. 65 (3), 471−487.

Lassaletta, L., Billen, G., Grizzetti, B., Anglade, J., Garnier, J., 2014. 50 year trends in nitrogen use efficiency of world cropping systems: the relationship between yield and nitrogen input to cropland. Environ. Res. Lett. 9 (10), 1−9.

Lassaletta, L., Billen, G., Garnier, J., Bouwman, L., Velazquez, E., Mueller, N.D., Gerber, J.S., 2016. Nitrogen use in the global food system: past trends and future trajectories of agronomic performance, pollution, trade, and dietary demand. Environ. Res. Lett. 11 (9), 095007.

Leip, A., Achermann, B., Billen, G., Bleeker, A., Bouwman, A.F., de Vries, A., Dragosits, U., Doring, U., Fernall, D., Geupel, M., Herolstab, J., Johnes, P., Le Gall, A.C., Monni, S., Neveceral, R., Orlandini, L., Prud'homme, M., Reuter, H.I., Simpson, D., Seufert, G., Spranger, T., Sutton, M.A., van Aardenne, J., Vos, M., Winiwarter, W., 2011. Integrating nitrogen fluxes at the European scale. In: Sutton, M.A., Howard, C.M., Erisman, J.W., Billen, G., Bleeker, A., Grennfelt, P., van Grinsven, H., Grizzetti, B. (Eds.), The European Nitrogen Assessment. Sources, Effects and Policy Perspectives. Cambridge University Press, Cambridge, pp. 345−376.

Maqsood, M.A., Awan, U.K., Aziz, T., Arshad, H., Ashraf, N., Ali, M., 2016. Nitrogen management in calcareous soils: problems and solutions. Pakistan J. Agric. Sci. 53 (1), 79−95.

McLauchlan, K.K., Williams, J.J., Craine, J.M., Jeffers, E.S., 2013. Changes in global nitrogen cycling during the Holocene epoch. Nature 495 (7441), 352−355.

McNeill, A., Unkovich, M., 2007. The nitrogen cycle in terrestrial ecosystems. In: Marschner, P., Rengel, Z. (Eds.), Nutrient Cycling in Terrestrial Ecosystems, Soil Biology, vol. 10. Springer, Berlin, Heidelberg. https://doi.org/10.1007/978-3-540-68027-7_2.

Naeem, M., Khan, K., Rehman, S., Iqbal, J., 2007. Environmental assessment of ground water quality of Lahore area, Punjab, Pakistan. J. Appl. Sci. 7 (1), 41−46.

Norton, J.M., 2008. Nitrification in agricultural soils. Nitrog. Agric. Syst. 49, 173−199.

Oenema, O., Tamminga, S., 2005. Nitrogen in global animal production and management options for improving nitrogen use efficiency. Sci. China, Ser. A Life Sci. 48 (2), 871−887.

Oenema, O., Bannink, A., Sommer, S., Velthof, G.L., 2001. Gaseous nitrogen emissions from livestock faming systems. In: Follett, R., Hatfield, J. (Eds.), Nitrogen in the Environment: Sources, Problems and Management. Elsevier, Amsterdam, pp. 255–289.

Pan, B., Lam, S.K., Mosier, A., Luo, Y., Chen, D., 2016. Ammonia volatilization from synthetic fertilizers and its mitigation strategies: a global synthesis. Agric. Ecosyst. Environ. 232 (16), 283–289.

PCRWR, 2005. Annual Report 2005-6 Part 2. Pakistan Council for Research in Water Resources PCRWR, Islamabad, Pakistan, 2008a. www.pcrwr.gov.pk/Annual%20Reports/New%20Annual%20Repot%202005-06_2.pdf.

Phillips, S.M., 2012. Dietary protein requirements and adaptive advantages in athletes. Br. J. Nutr. 108 (S2), 158–167.

Podgorski, J.E., Eqani, S., Khanam, T., Ullah, R., Shen, H., Berg, M., 2017. Extensive arsenic contamination in high-pH unconfined aquifers in the Indus Valley. Sci. Adv. 3 (8), 1–10.

Raza, S., Zhou, J., Aziz, T., Afzal, M.R., Ahmed, M., Javaid, S., Chen, Z., 2018. Piling up reactive nitrogen and declining nitrogen use efficiency in Pakistan: a challenge not challenged (1961–2013). Environ. Res. Lett. 13 (3), 034012.

Raza, S., Chen, Z., Ahmed, M., Afzal, M.R., Aziz, T., Zhou, J., 2019. Dicyandiamide application improved nitrogen use efficiency and decreased nitrogen losses in wheat-maize crop rotation in Loess Plateau. Arch. Agron Soil Sci. 65 (4), 450–464.

Raza, S., Miao, N., Wang, P., Ju, X., Chen, Z., Zhou, J., Kuzyakov, Y., 2020. Dramatic loss of inorganic carbon by nitrogen-induced soil acidification in Chinese croplands. Global Change Biol. 26 (6), 3738–3751.

Riddick, S., Ward, D., Hess, P., Mahowald, N., Massad, R., Holland, E., 2016. Estimate of changes in agricultural terrestrial nitrogen pathways and ammonia emissions from 1850 to present in the Community Earth System Model. Biogeosciences 13 (11), 3397–3426.

Sailesh, N., Behera, S.N., Sharma, M., Aneja, V.P., Balasubramanian, R., 2013. Ammonia in the atmosphere: a review on emission sources, atmospheric chemistry and deposition on terrestrial bodies. Environ. Sci. Pollut. Res. 20, 8092–8131.

Sanz-Cobena, A., Misselbrook, T., Camp, V., Vallejo, A., 2011. Effect of water addition and the urease inhibitor NBPT on the abatement of ammonia emission from surface applied urea. Atmos. Environ. 45 (8), 1517–1524.

Science Advisory Board, 2011. Reactive Nitrogen in the United States: An Analysis of Inputs, Flows, Consequences, and Management Options Office of the U.S. EPA Administrator Washington, DC.

Seitzinger, S., Harrison, J.A., Böhlke, J.K., Bouwman, A.F., Lowrance, R., Peterson, B., Drecht, G.V., 2006. Denitrification across landscapes and waterscapes: a synthesis. Ecol. Appl. 16 (6), 2064–2090.

Šimek, M., Linda, J., David, W.H., 2002. What is the so-called optimum pH for denitrification in soil? Soil Biol. Biochem. 34 (9), 1227–1234.

Singh, J., Kunhikrishnan, A., Bolan, N.S., Sagar, S., 2013. Impact of urease inhibitor on ammonia and nitrous oxide emissions from temperate pasture soil cores receiving urea fertilizer and cattle urine. Sci. Total Environ. 465 (1), 56–63.

Stocker, T. (Ed.), 2014. Climate Change 2013: The Physical Science Basis: Working Group I Contribution to the Fifth Assessment Report of the Intergovernmental Panel on Climate Change. Cambridge university press.

Smil, V., 2002. N and food production: proteins for humans' diets. Ambio 31 (2), 126–131.

Soares, J.R., Cantarella, H., de Campos, M.M.L., 2012. Ammonia volatilization losses from surface-applied urea with urease and nitrification inhibitors. Soil Biol. Biochem. 52, 82–89.

Steinfeld, H., Gerber, P., Wassenaar, T., Castel, V., Rosales, M., de Haan, C., 2006. Livestock's Long Shadow. Environmental Issues and Options. Food and Agriculture Organization of the United Nations, Rome.

Syakila, A., Kroeze, C., 2011. The global nitrogen budget revisited. Greenhouse Gas Meas. Manag. 1, 17–26.

Tahir, M.A., Rasheed, H., 2008. Distribution of nitrate in the water resources of Pakistan. Afr. J. Environ. Sci. Technol. 2 (11), 397–403.

Tian, H., Xu, R., Canadell, J.G., Thompson, R.L., Winiwarter, W., Suntharalingam, P., et al., 2020. A comprehensive quantification of global nitrous oxide sources and sinks. Nature 586 (7828), 248–256.

Van der Stelt, B., Temminghoff, E.J.M., Van Vliet, P.C.J., Van Riemsdijk, W.H., 2007. Volatilization of ammonia from manure as affected by manure additives, temperature and mixing. Bioresour. Technol. 98 (18), 3449–3455.

Van Meter, K.J., Basu, N.B., Veenstra, J.J., Burras, C.L., 2016. The nitrogen legacy: emerging evidence of nitrogen accumulation in anthropogenic landscapes. Environ. Res. Lett. 11 (3), 035014.

Wang, S., Nan, J., Shi, C., Fu, Q., Gao, S., Wang, D., Cui, H., Saiz-Lopez, A., Zhou, B., 2015. Atmospheric ammonia and its impacts on regional air quality over the megacity of Shanghai, China. Sci. Rep. 5 (1), 1–13.

Wang, M., Tang, T., Burek, P., Havlík, P., Krisztin, T., Kroeze, C., et al., 2019. Increasing nitrogen export to sea: a scenario analysis for the Indus River. Sci. Total Environ. 694, 133629.

WHO, 2007. World Health Organization. Protein and Amino Acid Requirements in Human Nutrition WHO Technical Report Series Number 935.

Wirsenius, S., Azar, C., Berndes, G., 2010. How much land is needed for global food production under scenarios of dietary changes and livestock productivity increases in 2030? Agric. Syst. 103 (9), 621–638.

Wood, S., Henao, J., Rosegrant, M., 2004. The role of nitrogen in sustaining food production and estimating future nitrogen fertilizer needs to meet food demand. In: Agriculture and the Nitrogen Cycle: Assessing the Impacts of Fertilizer Use on Food Production and the Environment, pp. 245–265.

Xuejun, L., Fusuo, Z., 2011. Nitrogen fertilizer induced greenhouse gas emissions in China. Curr. Opin. Environ. Sustain. 3 (5), 407–413.

Yan, X., Chaopu, T., Peter, V., Deli, C., Adrian, L., Zucong, C., Zhaoliang, Z., 2014. Fertilizer nitrogen recovery efficiencies in crop production systems of China with and without consideration of the residual effect of nitrogen. Environ. Res. Lett. 9, 095002.

Zaehle, S., 2013. Terrestrial nitrogen–carbon cycle interactions at the global scale. Phil. Trans. Biol. Sci. 368 (1621), 20130125.

Zhang, W.F., Dou, Z.X., He, P., Ju, X.T., Powlson, D., Chadwick, D., Zhang, F.S., 2013. New technologies reduce greenhouse gas emissions from nitrogenous fertilizer in China. Proc. Natl. Acad. Sci. U.S.A. 110 (21), 8375–8380.

Zhang, X., Davidson, E.A., Mauzerall, D.L., Searchinger, T.D., Dumas, P., Shen, Y., 2015. Managing nitrogen for sustainable development. Nature 528 (7580), 51–59.

Zhang, B., Tian, H., Lu, C., Dangal, S.R., Yang, J., Pan, S., 2017. Global manure nitrogen production and application in cropland during 1860–2014: a 5 arcmin gridded global dataset for Earth system modeling. Earth Syst. Sci. Data 9 (2), 667–678.

Zhang, X., Zou, T., Lassaletta, L., Mueller, N.D., Tubiello, F.N., Lisk, M.D., et al., 2021. Quantification of global and national nitrogen budgets for crop production. Nat. Food 2, 529–540.

---

## Further reading

David, F., Mhairi, C., Ute, S., Mark, S.A., Neil, C.J., Stefan, R., Lucy, S.J., Alan, J., Bruna, G., James, G.N., Peter, V., Allison, L., Alexander, B.F., Klaus, B., Frank, D., David, S., Marcus, A., Maren, V., 2013. The global nitrogen cycle in the twenty-first century. Phil. Trans. Biol. Sci. 368 (1621), 1–13.

Galloway, J.N., Dentener, F., Burke, M., Dumont, E., Bouwman, L., Kohn, R., Mooney, L., Seitzinger, S.P., Kroeze, C., 2010. The impacts of animal production systems on the nitrogen cycle. In: Steinfeld, H., Mooney, H.A., Schneider, F., Neville, L.E. (Eds.), Livestock in a Changing Landscape. Drivers, Consequences and Responses. Island Press, Washington, DC, pp. 83–95.

# Drivers of increased nitrogen use in Pakistan

**Masood Iqbal Awan[1], Sajjad Raza[2], Amara Farooq[3], Allah Nawaz[3,4], Tariq Aziz[1]**

[1]*University of Agriculture Faisalabad, Sub-Campus at Depalpur, Okara, Punjab, Pakistan;* [2]*School of Geographical Sciences, Nanjing University of Information Science & Technology, Nanjing, Jiangsu, China;* [3]*Institute of Soil and Environmental Sciences, University of Agriculture Faisalabad, Punjab, Pakistan;* [4]*Soil Chemistry Section, Ayub Agricultural Research Institute, Faisalabad, Punjab, Pakistan*

## 4.1 Introduction

The ever-increasing population has evidently disturbed the balance of food demand and supply, and the gap is increasing particularly in developing countries. The world's need for food and fiber will increase as the population is anticipated to rise by about 35% from the current 6.9 billion to around 10 billion in 2050 (Gilland, 2002; Sharma and Bali, 2017). The past gains in agricultural productivity were accomplished with the introduction of high yielding crop varieties, improved agronomic practices, and expansion of the irrigated area during the Green Revolution era of 1960s. Compared to the traditional varieties, the modern bioengineered varieties were more responsive to inputs, in particular chemical fertilizers in terms of economic yields and lodging resistance. Cereal production increased more than three times due to gains in both productivity and the irrigated area with a concurrent manifold increase in the use of chemical fertilizers especially those supplying nitrogen (N). For example, mineral N consumption increased from 11.8 Tg in 1961 to 108 Tg in 2017 (Food and Agriculture Organization, 2020).

Population projections for 2050 necessitate that we should increase our agricultural production by as much as 70%. In order to meet the target for enhanced crop production, N consumption will also increase by 2.7 times (Hu et al., 2020; Rahman and Zhang, 2018; Ti et al., 2019; Zhang et al., 2013). Mineral N fertilizers contribute an increase of approximately 50% in food production, helping in decreased levels of world's hunger (Norton and Ouyang, 2019; Yang et al., 2016, Fig. 4.1); hence it is directly related to the Sustainable Development Goals (SDG2), i.e., Zero Hunger. Nonetheless, approximately 50% of the N applied in an agro-ecosystem is lost due to various N transformations in the soil (Cavigelli et al., 2012). High N use also means high N losses owing to low N use efficiency around the world (Hu et al., 2020) particularly in South Asia (Addy et al., 2020; Raza et al., 2018). Thus, there is need to focus on more rational use of N as compared to the past traditional practices because N has a big cost in terms of investment and environmental

Nitrogen Assessment. https://doi.org/10.1016/B978-0-12-824417-3.00004-6

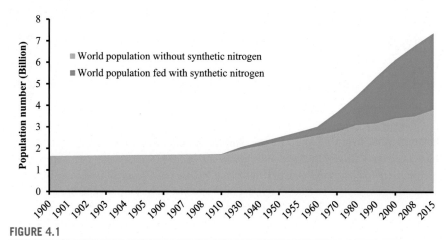

**FIGURE 4.1**

Global population supported by synthetic N fertilizer application.

*Based on Erisman, J.W., Sutton, M.A., Galloway, J., Klimont, Z., Winiwarter, W., 2008. How a century of ammonia synthesis changed the world. Nat. Geosci. 1 (10), 636–639; Stewart, W.M., Dibb, D.W., Johnston, A.E., Smyth, T.J., 2005. The contribution of commercial fertilizer nutrients to food production. Agron. J. 97 (1), 1–6.*

consequences (Blum, 2013; Cirera and Masset, 2010; Godfray et al., 2010; Kanter et al., 2020). Already, irrational use of N due to overdosing or low use-efficiency narrowed our net benefits along with devastating effects on the environment, e.g., soil, air, and groundwater pollution, biodiversity reductions, eutrophication, and gaseous emissions. Societal cost of the devastating effects is rising risk level for human health due to increasing incidences of cancer and respiratory diseases (Clark and Tilman, 2008; Diaz and Rosenberg, 2008; Mosier et al., 2004; Townsend et al., 2003).

Another challenge is increased urbanization, which affects land use options with a shift in dietary patterns and energy consumption (Ishaque, 2017; Shirazi, 2012). With increasing urban population and prosperity, our dietary patterns are changing in favor of more animal-based food. The share of animal-based food N to total food N is significantly higher in urban diets than in rural diets, which can potentially cause health and environmental risks (Food and Agriculture Organization, 2020; Tilman and Clark, 2014).

With the increased urbanization, demand for transport vehicles, industrial products, and energy consumption has also risen in the past and will continue so in the future. Resultantly, more gaseous emissions from vehicles, industries, and power houses are further polluting the environment. Land use patterns are changing to meet the requirements of residential sector and need to produce more food from diminishing resources that warrants increased use of N in agriculture (Bhalli et al., 2012) and its release back to environment polluting air and water resources thus decreasing biodiversity.

Factors with causal relations for changes in N fixation or application are called as "N drivers." For instance, changes in population number (i.e., primary driver) over time will influence food production (i.e., secondary driver). The extent of changes in an economic activity like agriculture is often reflected by the parameters underlying that activity. Exogenous drivers such as demography, changing dietary patterns, and technological advances must be linked with policy actions in agriculture, trade, and environmental sectors. An understanding about human activities and their correlation with N drivers at local, regional, and global scales is essential for rational management of N (Lassaletta et al., 2014). In this chapter, the correlation between rising population and various sectors including agriculture, industry, transport, and power with a particular focus on data from Pakistan has been synthesized.

## 4.2 Population

Being a structural component of proteins, N is crucial to sustainable human nutrition and is an integral part of daily diet. For humans, the recommended daily protein intake for adults is $0.8-0.9$ g $kg^{-1}$ of body weight (Phillips, 2012) and recommended per capita annual N intake is $2.5-3.5$ kg N $yr^{-1}$ (World Health Organization, 2007).

By 2050, the human population will expectedly reach 10 billion with major increases are being projected for regions in Africa and Asia, e.g., South Asia is the most populous region with 25% global population. Another associated challenge is that urban population is increasing more rapidly than the rural population due to urbanization (Food and Agriculture Organization, 2020). One major cause of increasing trend in the human population is the use of N fertilizers owing to its role in ensuring food security by increasing crop production. Without N use, there would have been 3.5 billion less mouths to feed than the current 7.8 billion (Zhang et al., 2013, Fig. 4.1). As most of the countries in South Asia and Africa are developing or underdeveloped with majority of the population living at or below the line of poverty, food security will remain a major challenge for these regions in particular and for the world in general. In this regard, N fertilization will continue to play a decisive role in populous countries like Pakistan, the fifth-most populous country in the world with 220.9 million in 2020 (Fig. 4.2), which is expected to reach 313.97 million people by 2100 (Ahmad and Farooq, 2010; Food and Agriculture Organization, 2020; Pakistan Bureau of Statistics, 2018). Nonetheless, major challenge or uncertainty in such population forecasts was migration across the borders, which is further complicated by the post-COVID-19 situation causing back migration of people to native countries. Agricultural production and productivity must increase to meet food, fiber, and fuel demands for population increasing at an alarming rate. In order to feed the growing population, productivity of wheat—top ranked staple grain—must increase from the current 2.8 to 3.8 Mg $ha^{-1}$ by 2030, which is a huge challenge for resource-constrained farmers (Ali, 2018).

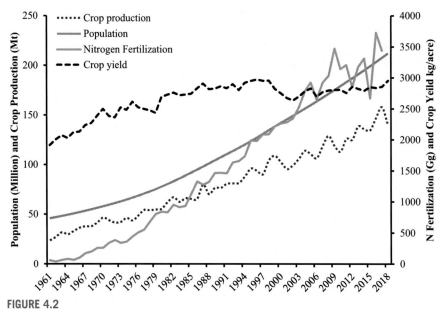

**FIGURE 4.2**

Changes in population number, N fertilization amounts, crop yields, and crop production in Pakistan during 1961–2018.

*Based on Food and Agriculture Organization, 2020. FAOSTAT Database on Agriculture. Food and Agriculture Organization of the United Nations (FAO), Rome, Italy. www.fao.org/faostat/en/#data. (Accessed 10.12.2020).*

## 4.3 Food crop production

Plants are primary producers, on which human and livestock depend for their nutritional requirements. Agricultural systems aim at increasing plant production under field conditions. During the past five decades, approximately 40% increase in per capita food production is linked solely to N fertilization (Brown, 1999; Smil, 2002). Pakistan has recorded substantial increases in field crop production and per acre yield during the past six decades (Fig. 4.2). The crop production steadily increased from 24 MT in 1961 to 158 MT in 2017; the per acre yield increased from 1.91 to 2.97 Mg during the same period (Food and Agriculture Organization, 2020). Such a drastic increase in productivity was mainly achieved with intensive use of inputs, in particular mineral N fertilizers.

Most soils in Pakistan (>95%) are low in organic matter and are essentially N-deficient. Considering the population growth rate (~2% per annum) and diminishing land availability for agriculture necessitates our local agricultural systems to produce more from less—more food from less resources of land, water, and energy. Increased production per unit area demands a simultaneous rise is N fertilization. Moreover, we need cultivars with high nutrient capture and utilization efficiencies.

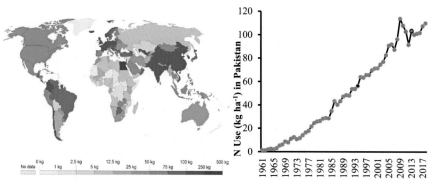

**FIGURE 4.3**

Nitrogen fertilizer use per hectare of cropland during 2017 in the world (left) and changes in N fertilizer use (kg ha$^{-1}$) during 1960−2017 in Pakistan (right).

*Adapted from Ouworldindata.org/Charts; Food and Agriculture Organization, 2020. FAOSTAT Database on Agriculture. Food and Agriculture Organization of the United Nations (FAO), Rome, Italy. www.fao.org/faostat/en/#data. (Accessed 10.12.2020).*

During the last six decades, Pakistan has experienced a massive increase (65-fold) in fertilization (Food and Agriculture Organization, 2020; Raza et al., 2018). In 1960−61, the fertilizer consumption in Pakistan was 73 Gg yr$^{-1}$, which increased to 1080 Gg yr$^{-1}$ in 1980−81, 2963 Gg yr$^{-1}$ in 1999−2000, and 4768 Gg yr$^{-1}$ in 2017−18 (Food and Agriculture Organization, 2020). The share of N fertilizer is around 80% of the total nutrients consumed in the country. N use in Pakistan is about 3% of the global N use (Food and Agriculture Organization, 2020). Per unit land area, N use in Pakistan is more than 150 kg ha$^{-1}$ (Raza et al., 2018; Vitousek et al., 2009), which is fairly higher than many of the countries across Asia, Europe, Africa, and Australia (Fig. 4.3).

Although increase in N fertilizers helped increase crop production but NUE is far below than the global average. In addition, about 3.6 Tg surplus N (175 kg N ha$^{-1}$ yr$^{-1}$) is being added up annually in the atmosphere or water (Raza et al., 2018). Considering the recent estimates for population in 2050, the use of N fertilizers would increase up to 5.9 Tg yr$^{-1}$ with a rise in surplus N of up to 4.4 Tg if it is not rationalized. Sustainable and efficient use of N fertilization should be top priority for Pakistan to fulfill food demand and keep environmental pollution as well as crop production costs at minimum.

## 4.4 Feed crop production

Livestock contributes about 56.3% of the value added in agriculture, and nearly 11% to the agricultural gross domestic product (Pakistan Bureau of Statistics, 2018). To meet the rising demand for meat and milk, a significant increase in livestock and poultry is required. Like crop production, animal stock also increased steadily to sustain the rising population number in Pakistan (Fig. 4.2; Fig. 4.4). Cattle, buffalo, sheep, goats, and poultry are the main producers of milk and/or meat in Pakistan and current animal stock reached 197 million, substantially higher than 42 million heads in 1961 (Food and Agriculture Organization, 2020). During the last six decades, livestock sector has increased by 370% and poultry stocks have increased by about 44 times (Fig. 4.4). In 1961, Pakistan had only 21 million heads of cattle and buffalo and 19 million heads of sheep and goats, which have increased to 85 million heads and 105 million heads, respectively. Poultry sector was less developed and showed slower progress during 1960—90 (12 million heads to 82 million heads) and afterward an exponential increase from 82 million heads to 528 million heads was recorded during 1990—2018 due to the involvement of private sector and affordable consumer prices. In fact, poultry sector is shifting the main burden of meat requirements from livestock sector in Pakistan. Intensification of livestock production leads to increased use of land for growing feedstuffs and to a rise in the use of fertilizers.

A huge increase during the past 3—4 decades in livestock population has resulted in a significant increase in feed crop production. About one-third of the global cereal production is fed to animals (Food and Agriculture Organization, 2020). Considering the expected rise in livestock population (359 million heads in 2100) and poultry birds, the crop production needs to be enhanced substantially and this can only be achieved by best management practices including fertilization. Hence, a significant increase in N use is expected in future for animal sector in addition to fulfilling the human needs. Secondly, the use efficiency of N is seriously low in livestock and poultry sector in comparison with crop sector. The conversion efficiency of plant into animal matter is ~10%; this means more people could be supported from the same amount of land if they use plant-derived food. Hence, livestock

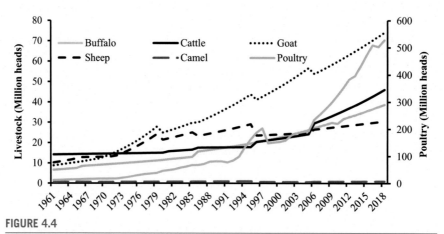

**FIGURE 4.4**

Changes in population number of livestock and poultry in Pakistan during 1961—2018.

*Reproduced from Food and Agriculture Organization, 2020. FAOSTAT Database on Agriculture. Food and Agriculture Organization of the United Nations (FAO), Rome, Italy. www.fao.org/faostat/en/#data. (Accessed 10.12.2020).*

sector is a significant driver of increased N use and subsequently increase in N surplus, which needs to be managed for better air, water, and soil quality.

In order to increase and sustain enhanced food availability, livestock and dairy sectors were encouraged by Pakistan's government. During the past six decades, averaged per capita animal availability was one to fulfill the milk and meat requirements (Food and Agriculture Organization, 2020). Considering this, the combined population of buffalo, cattle, sheep, and goat should increase from 197 million heads current to 301 million heads in 2050 and up to 359 million heads in 2100 in Pakistan (Food and Agriculture Organization, 2020, Fig. 4.4). If the required increase in animal production is not achieved, it will put immense pressure on crop sector for fulfilling needs.

## 4.5 Dietary patterns

Typically, in Asian countries, with improved economy and lifestyle in major cities, a significant change in dietary patterns has been observed. The growing trend in livestock population is related not only to an increase in the human population but also to an increase in the share of animal products in the diet (Raza et al., 2018). With rise in income levels, there is a gradual shift in dietary pattern from crop-based diets/protein sources to meat-based protein sources. Statistics showed that diet patterns are changing with a gradual decrease in the share of protein derived from crop sources (from 72% to 58%) and an increase in the share of animal protein (from 27% to 41%) (Fig. 4.5). Total meat production in Pakistan has increased from 0.36 million tons to 3.87 million tons in 2018.

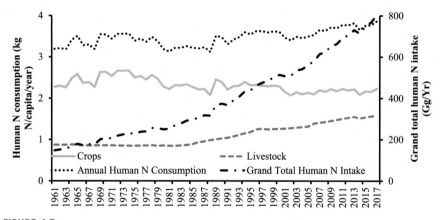

**FIGURE 4.5**

Changes in human N intake from different food sources in Pakistan during 1961–2017.

*Reproduced from Food and Agriculture Organization, 2020. FAOSTAT Database on Agriculture. Food and Agriculture Organization of the United Nations (FAO), Rome, Italy. www.fao.org/faostat/en/#data. (Accessed 10.12.2020).*

During the past five decades, Pakistan has experienced a 14% increase in total protein consumption, with the share of animal protein increasing by 48%. Currently, every individual in Pakistan consumes around 24 kg protein per year, equivalent to 3.87 kg N cap$^{-1}$ yr$^{-1}$, which is within the range of recommended healthy diet (Fig. 4.5). The total protein consumption has increased by 12% from 19.95 kg yr$^{-1}$ in 1961 to 24.18 kg yr$^{-1}$ in 2017 (Food and Agriculture Organization, 2020). Combined with the increase in population, this increase in the share of animal protein has caused a substantial increase in the consumption of animal products.

Rising livestock population means more usage of grain and fodder crops as their feed (as discussed in Section 4.3), which will result in their decreased availability for human food consumption on one hand and increased N fertilizer use on the other. The inherently inefficient conversion of plant protein into animal protein makes meat responsible for a disproportionate share of environmental pressure (Gilland, 2002; Raney et al., 2009; Steinfeld et al., 2006). As a result of animal metabolism, 6 kg of plant protein is required to yield on average 1 kg of meat protein (Pimentel and Pimentel, 2003; Smil, 2000).

Consequently, a mere 15% of protein and energy in these crops will ever reach human mouths, and 85% are wasted (Kummu et al., 2012). Therefore, it is very important to keep protein consumption from animal sources at minimum considering the proportion of crops required for their feed as well as associated environmental issues like gaseous emissions.

## 4.6 Urbanization

Migration from rural areas to urban areas is a serious challenge around the globe in addition to international migrations. Urbanization is closely related to economic growth and people from rural areas move to big cities in search of job, business, and better living facilities.

Generally, cities have higher per capita income and more employment opportunities, so people from rural areas prefer to move to big cities for better livelihood opportunities. The rapid movement of people from rural to urban areas needs more resources and living spaces, and thus has a significant impact on land use. For example, large amounts of agricultural areas and forests are being replaced with urban and residential land. The changes in lifestyle (diet, domestic sanitation, transport use, energy use) and production (agriculture, industry, and services) has serious impacts on regional N flows and water environment (Liu et al., 2015).

Pakistan has the highest rate of migration from rural to urban area in South Asia. The urban population of the country is 37.2% as compared to 17.5% in 1950 (Food and Agriculture Organization, 2020, Fig. 4.6), while urban population of South Asia is 32.5%. As per UN population division estimates, urban population will be about 50% of the total population by 2025 in Pakistan.

Urbanization has serious impacts on natural resources, energy consumption, and environment. Moreover, the dietary patterns in cities are quite different than those in

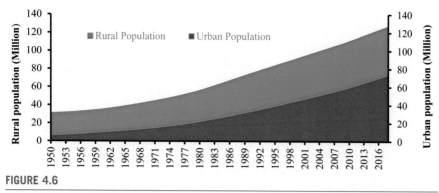

**FIGURE 4.6**

Population growth and urbanization in Pakistan from 1950 to 2018. Current Population is 220.9 million (Food and Agriculture Organization, 2020).

rural areas because of the better economic activities and opportunities. The dietary change from 1961 to 2013 has been discussed recently by Raza et al. (2018), which clearly indicate that the animal-based protein consumption increased significantly during this period and major shift has been observed during last 3 decades. The animal-based protein production has linear relation with livestock population and N utilization (see Section 4.4). Converting back to plant-based protein in diet would improve N use efficiency and reduce environmental costs of livestock production as well as reduce the land used for feed production and grazing.

Improved economic conditions in the cities also bring a significant change in energy requirement and per capita utilization. This increased energy demand is linearly related with higher use of fossil fuels and resultantly N emissions. Secondly, increased urbanization is also linked with higher use of transport either private or public, thus fuel consumption has been increased significantly during the past two decades.

## 4.7 Land use

Global agricultural land use both for food and feed production is 4.13 billion hectares which can be further subdivided into cropland (1.24 billion ha) and pastures (2.89 billion ha). The cropland can further be divided into area used for human food (740 Mha) and for animal feed (538 Mha) (Food and Agriculture Organization, 2017; Ourworldindata.org/Charts). Land use is an important factor regulating the greenhouse gas emissions. Out of total greenhouse gas emissions, 26% are released from food sector, while 74% from nonfood sectors (Poore and Nemecke, 2018).

Land resources are finite and arable lands are squeezing because of desertification, urbanization, and pollution. As land is a major resource for crop production, hence agricultural land use and fertilizer N input are strongly affected by factors such as income of farmers, prices of agricultural products and fertilizers, policies of land use, trend in urbanization, and population growth and trade. Pakistan's total land area is 79.6 million hectares and about 22 million hectares (23% of total land area) is cultivated. Out of this, 18 million hectares are irrigated and 4 million hectares are rainfed.

Cropped area (area under cereals/fiber/oilseed/pulse crops, fruits, vegetables, roots, tubers, etc.) was only 12.7 million hectares in 1961 which increased gradually till 2000 and reached 23.45 million hectares (including area sown more than once) in Pakistan. After 2000 till now, the area under crops has not increased and remained constant at around 23 million hectares.

Land use affects N fertilization in several ways in Pakistan. Some of them are discussed below:

1. Pakistani population is increasing exponentially, but the area under crop cultivation has not increased at all since the last two decades. In fact, per capita cropland has gradually decreased from 0.27 ha in 1961 to only 0.1 ha in 2017. This is putting immense pressure to produce more food from the same or decreased land area and is by far a leading cause to an increased use of N fertilizers in Pakistan.

2. With intensification approach, the cropping intensity increased in many parts of the world including Pakistan. Double cropping (cotton/rice followed by wheat) is a common practice in most parts of the country with some regions practicing triple cropping. Intensified cropping systems are major drivers of increased use of inputs like fertilizers, pesticides, and irrigation. Taking two to three crops every year from the same field feeds heavily on the nutrients and depletes soil capability. Getting higher yields from such soils requires continuous application of fertilizers in higher amounts. This is an important driver which accelerates the use of N fertilizers.

3. Intensity of N fertilizers use also depends on the quality of land. Soils that are fertile and productive can produce higher yield from less doses of N fertilizers. On the other hand, soils which are degraded and are deficient in nutrients require large amount of N fertilizers to produce higher yields. Pakistan faces serious issues of land degradation from waterlogging, salinity, nutrient mining, and soil erosion. Pakistan is a typical country where soil degradation and lower soil organic matter are major threats contributing toward lower crop productivity. Around 95% of the Pakistani land is deficient in organic matter, which will intensify the use of fertilizers and irrigation (Niazi, 2003). This is another aspect, which has caused exponential increase in the consumption of N fertilizers. Pakistan has, however, been overharvesting the one-fourth of its land that is relatively more productive.

4. Cash crops such as cotton, rice, maize, wheat, potato, sugarcane, tobacco, oilseeds, and certain fruits offer more income and therefore represent an important driver of N fertilization in Pakistan. Farmers spend more money on inputs such as N fertilizers to get higher yields. Further, cash crops are usually grown by progressive growers who have more resources, which further allow them to apply more fertilizers.

**5.** Landholding is another major problem of Pakistan which affects application of inputs like N fertilizers. Large landholders (i.e., holding more than 50 acres in Punjab/Khyber Pakhtunkhwa and >64 acres in Sindh and Balochistan) are engaged in commercial farming that is highly input intensive, especially intensity of fertilizers, irrigation, and pesticides on crops such as cotton. Large landholders have enough sources and income to buy expensive fertilizers to support their cash crops and profitability. Similarly, small landholders (i.e., up to 12.5 acres in Punjab/Khyber Pakhtunkhwa, 16 acres in Sindh, and 32 acres in Balochistan) are engaged in subsistence farming to survive on the limited land that they tend to apply fertilizers intensively to reap enough harvest for self-subsistence on one hand, and compensation for high cost of production and rent in the land (in the case of tenants) on the other hand. Small landholders, having limited access to such resources, tend to intensify their use for subsistence. Land use practices of large and small holders are unsustainable, which culminate in nonjudicious and irrational approaches for N fertilization.

Among crops, cereals (wheat, maize, rice, sorghum, millets, and barley) occupy more than 65% of total cropped area in Pakistan (Fig. 4.7). Around 60% of the fertilizer is applied to cereals in Pakistan, with wheat alone consuming 43% due to its large acreage of around 9 million hectares. Fiber crops which include cash crop like cotton also consume around about 20% of fertilizers and the share of sugar crops is around 8%. The share of vegetables is around 2% and of fruits is about 4% in total

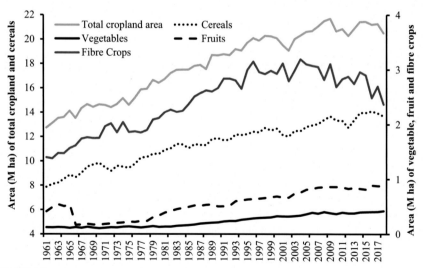

**FIGURE 4.7**

Changes in crop area (million hectare) under different crops in Pakistan during 1961–2018.

*Based on Food and Agriculture Organization, 2020. FAOSTAT Database on Agriculture. Food and Agriculture Organization of the United Nations (FAO), Rome, Italy. www.fao.org/faostat/en/#data. (Accessed 10.12.2020).*

cropland area and their combined share of fertilizer consumption is around 5.5% (Fig. 4.7). Forests, both natural and man-made, which cover only about 4% of the total land area, are not usually supplied with fertilizers. Similarly, grasslands and pastures are rarely fertilized in Pakistan.

## 4.8 Energy sector

Burning of fossil fuels is directly related to N emissions. The industrialization and urbanization have resulted in a significant increase in energy consumption. Industrial activities such as building, ceramics, bricks making, smelting, cement manufacturing, and mining along with the generation of electricity and burning of biomass are resulting in air-polluting emissions (Alam et al., 2015; Ali and Athar, 2010; Shah et al., 2012).

Pakistan's share of greenhouse gas emissions is almost negligible in the global emissions with a total share of less than 1% reported in 2019 (Food and Agriculture Organization, 2020). The current energy demand in Pakistan is nearly 25,000 MW, while production is about 22,000 MW (Government of Pakistan, 2020). The government has set up several power plants in to fill the demand—supply gap during the past 2 decades, which increased installed capacity of electricity from 22,860 MW in 2012 to 35,972 MW in 2020 (Government of Pakistan, 2020). During the last 2 decades, because of severe shortfall of energy, the usage of generators and small power units in the industry was tremendously increased resulting in increased utilization of gasoline and compressed natural gas (CNG).

At present Pakistan is consuming 65.3 million tons of oil equivalent of primary energy. It is expected to rise primary energy consumption to 147 million tons oil equivalent till 2030. Almost 68% of energy demand is fulfilled from the domestic resources, principally natural gas. The coal consumption has been increased since 2017 as new power generation plants have been installed. The energy mix of Pakistan during 2000—19 indicates that major share is of oil, hydro, and gas, while nuclear, solar, wind, and renewable sources have minor share in total energy mix (Fig. 4.8).

It is appropriate to cite government plans to adopt coal-based technologies to generate electricity in the coming years due to the prevailing present energy crisis. Coal-fired power projects like Sahiwal coal power project, add electricity to the national grid. Nevertheless, the plants will also add to rising emissions. Coal-fired power plants are more prone to increase both primary pollution (direct) as well as secondary pollution such as particulate matter (gas to particle conversion). Likewise, coal combustion primarily depends on several factors such as technology used for the combustion process, quality of the coal, and the subsequent climatic conditions (Galloway et al., 2008; Jan et al., 2017; Nayyar et al., 2014; Saeed et al., 2015; Sheikh, 2003; Tahir et al., 2010).

With the increased energy demands, policy makers must focus on environment-friendly sources for power generation. Power generation from thermal power plants by burning of fossil fuels is a significant contributor to total GHG emissions. There

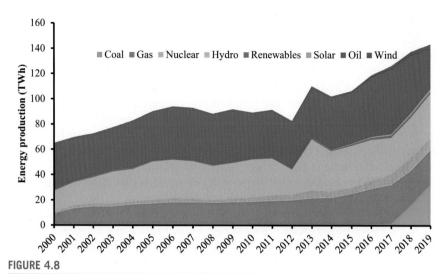

**FIGURE 4.8**

Energy production from different sources (TWh) in Pakistan from 2000 to 2019.

*Adapted from Ourworldindata.org.*

are a large number of thermal power plants, most of them independent power producers, currently operating in Pakistan, with more plants in the central and southern parts of the Punjab province (Ali et al., 2011; Irfan et al., 2015; Rehman et al., 2020). The total installed generation capacity of Pakistan as on June 30, 2020 was 38,719 MW that includes thermal, wind, solar, hydroelectric, bagasse, nuclear, and small power producers/captive power plants (National Electric Power Regulatory Authority, 2020).

Environmental hazards associated with coal-fired power plants should be accounted for intriguing of government programs to integrate energy generation technologies. Government and policymakers must work to make important decisions regarding the country's fuel transformation as well as future fuel economy. According to an estimate, 70% of rural people are fulfilling energy needs by using biomass and firewood. Daily oil consumption in Pakistan was reported at 445,965 barrels during 2019, whereas petroleum consumption is expected to increase from 20,177 MT by 2012 to 53,079 MT by 2030. Transport and power sector are major consumers of petroleum (oil) and there is an increasing trend from 2000 to 2020 in both the sectors as compared to other sectors such as industry and agriculture (Rehman and Deyuan, 2018).

## 4.9 Transport sector

Globally, the transport sector plays a major role in N oxides (NOx) emissions especially from road transport that accounts for about 70% of NOx emissions and about

30% of particulate matter to the environment. Combustion of fuel in automobiles/ vehicles is a big source of NOx emissions causing air pollution. An increasing trend for urban population combined with higher per capita income has resulted in an increased number of vehicles in Pakistan. The number of vehicles has reached 25.98 million in 2020 (Fig. 4.9) as compared to 2.7 million in 1990 (Government of Pakistan, 2020). Improved road infrastructure during the last 15 years has also contributed toward more mileage of vehicles per year. High-temperature combustion processes that occur in automobile engines are major sources of $NO_2$ from anthropogenic sources (Mauzerall et al., 2005), hence NOx emission increased substantially during these two decades in Pakistan (Raza et al., 2018). Fuel sources being consumed in Pakistan include gasoline, diesel, CNG, or liquefied petroleum gas, and jet fuel. Petroleum consumption by different sectors includes 2.9% by the households, 11.0% by the industry, 1.5% by the agriculture, 51.9% by the transport, and 32.8% by the power sector. This increase is due to the rising number of motorcycles at the rate of 15% in the last 5 years. Three-wheelers vehicles currently use gasoline and liquefied petroleum gas (70:30). Due to the shortage of CNG, four wheelers vehicles mainly use petrol. Demand for the natural gas is expected to increase to 13.29 cubic meters per day for the next 10 years till 2030. There would be a shortfall of about 15 billion cubic feet per day by the year 2030. Pakistan's natural gas reserves are dwindling and if the current gas situation exists, Pakistan will face severe gas shortages.

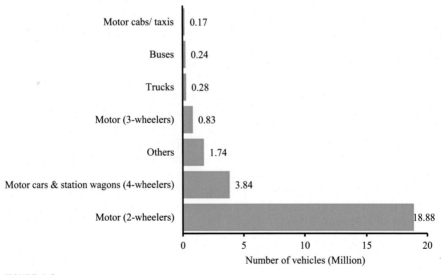

**FIGURE 4.9**

Number of different vehicles in Pakistan as of year 2020.

*Based on Government of Pakistan, 2020. Economic Survey of Pakistan, 2019-20. Economic Advisory Wing, Finance Division, Government of Pakistan, Islamabad.*

## 4.10 **Industry**

Besides fertilizers, Haber—Bosch N is widely used in industrial products like nylon. Gu et al. (2013) calculated that the industrial N flux increased globally from 2.5 in 1960 to 25.4 Tg N yr$^{-1}$in 2008, which is comparable to the NOx emissions from fossil fuel combustion. In addition, more than 25% of industrial products, mainly nylon-based, tend to accumulate in human settlements due to their long service lives. Emerging reactive N species define new N-assimilation as well as decomposition pathways and change the way that reactive N is released into the environment and the losses of these reactive N species to the environment have grave human and ecosystem impacts. Incorporating industrial reactive N into urban environmental and biogeochemical models could help us advance the agenda of urban ecology and environment.

With a significant increase in gross domestic product during the last 5 decades, industrialization has increased in Pakistan. The increased industrialization is linked with high energy demand during the last 4—5 decades. Higher emissions from the industrial sector are linked with the burning of fuel for electricity generation and directly from factories producing N acids and fertilizers. The total production of N fertilizers in the country also increased from 2.2 million metric tons in 2002 to 3.2 million metric tons in 2019. Pakistan ranks eighth in N fertilizers production worldwide. Currently, 13 factories/plants are producing N fertilizers with a total annual production capacity of 8983 Gg (Government of Pakistan, 2020). Estimates suggest that industrial sector will consume around 49% energy supply by 2030.

## 4.11 **Conclusions**

Nitrogen consumption in Pakistan has been increased significantly and main increase was observed during the last 3 decades. Ever-increasing population is one of the main drivers for increased N utilization. The other drivers such as livestock production, urbanization, land use, energy consumption, and transport sector are linked directly with the population. There is need to produce more food (about 70%) to feed the estimated global population of about 10 billion in 2050. Moreover, urbanization has increased across the globe. In Pakistan, the rate of urbanization is the highest within South Asia. Urbanization with improved economic growth influences dietary patterns and requires more resources like energy, land, fertilizers, water, transport, etc. All such factors ultimately result in increased or increasing N use along with emissions; thus these must be considered for sustainable food production with minimum N footprints. Actionable policy recommendations require regular assessment of N dynamics across various sectors followed by strict regulatory measures.

# References

Addy, J.W.G., Ellis, R.H., Macdonald, A.J., Semenov, M.A., Mead, A., 2020. Investigating the effects of inter-annual weather variation (1968–2016) on the functional response of cereal grain yield to applied nitrogen, using data from the Rothamsted Long-Term Experiments. Agric. For. Meteorol. 284, 107898.

Ahmad, M., Farooq, U., 2010. The state of food security in Pakistan: future challenges and coping strategies. Pak. Dev. Rev. 49, 903–923.

Alam, K., Rahman, N., Khan, H.U., Haq, B.S., Rahman, S., 2015. Particulate matter and its source apportionment in Peshawar, Northern Pakistan. Aerosol. Air Qual. Res. 15, 634–647.

Ali, M.A., 2018. Status Paper on Wheat Crop in Pakistan. Plant Sciences Division, Pakistan Agriculture Research Council (PARC), Islamabad, Pakistan.

Ali, M., Athar, M., Khan, M.A., Niazi, S.B., 2011. Hazardous emissions from combustion of fossil fuel from thermal power plants based on turbine technologies. Hum. Ecol. Risk Assess. 17 (1), 219–235.

Ali, M., Athar, M., 2010. Impact of transport and industrial emissions on the ambient air quality of Lahore city, Pakistan. Environ. Monit. Assess. 171 (1–4), 353–363.

Bhalli, M.N., Ghaffar, A., Shirazi, S.A., Parveen, N., Anwar, M.M., 2012. Change detection analysis of land use by using geospatial techniques: a case study of Faisalabad Pakistan. Sci. Int. 24 (4), 539–546.

Blum, W.E., 2013. Soil and land resources for agricultural production: general trends and future scenarios-a worldwide perspective. Int. Soil Water Conserv. Res. 1 (3), 1–14.

Brown, L.R., 1999. Feeding nine billion. In: State of the World: A World Watch Institute Report on Progress toward a Sustainable Society. WW Norton & Company, New York, US, pp. 115–132.

Cavigelli, M.A., Grosso, S.J.D., Liebig, M.A., Snyder, C.S., Fixen, P.E., Venterea, R.T., et al., 2012. US agricultural nitrous oxide emissions: context, status, and trends. Front. Ecol. Environ. 10, 537–546.

Cirera, X., Masset, E., 2010. Income distribution trends and future food demand. Philos. T. R. Soc. B. 365, 2821–2834.

Clark, C.M., Tilman, D., 2008. Loss of plant species after chronic low-level nitrogen deposition to prairie grasslands. Nature 451 (7179), 712–715.

Diaz, R.J., Rosenberg, R., 2008. Spreading dead zones and consequences for marine ecosystems. Science 321 (5891), 926–929.

Erisman, J.W., Sutton, M.A., Galloway, J., Klimont, Z., Winiwarter, W., 2008. How a century of ammonia synthesis changed the world. Nat. Geosci. 1 (10), 636–639.

Food and Agriculture Organization, 2020. FAOSTAT Database on Agriculture. Food and Agriculture Organization of the United Nations (FAO), Rome, Italy. www.fao.org/faostat/en/#data. (Accessed 12 December 2020).

Food and Agriculture Organization, 2017. FAOSTAT: FAO Statistical Databases. Food and Agriculture Organization of the United Nations, Rome, Italy. http://www.fao.org/faostat/en/. (Accessed 12 December 2020).

Galloway, J.N., Townsend, A.R., Erisman, J.W., Bekunda, M., Cai, Z., Freney, J.R., et al., 2008. Transformation of the nitrogen cycle: recent trends, questions and potential solutions. Science 320 (5878), 889–892.

Gilland, B., 2002. World population and food supply: can food production keep pace with population growth in the next half-century? Food Pol. 27, 47−63.

Gu, B., Chang, J., Min, Y., Ge, Y., Zhu, Q., Galloway, et al., 2013. The role of industrial nitrogen in the global nitrogen biogeochemical cycle. Sci. Rep. 3, 2579.

Godfray, H.C.J., Beddington, J.R., Crute, I.R., Haddad, L., Lawrence, D., Muir, J.F., et al., 2010. Food security: the challenge of feeding 9 billion people. Science 327 (5967), 812−818.

Government of Pakistan, 2020. Economic Survey of Pakistan, 2019-20. Economic Advisory Wing, Finance Division, Government of Pakistan, Islamabad.

Hu, Y., Gabner, M.P., Weber, A., Schraml, M., Schmidhalter, U., 2020. Direct and indirect effects of urease and nitrification inhibitors on $N_2O$-N losses from urea fertilization to winter wheat in Southern Germany. Atmosphere 11, 782.

Irfan, M., Riaz, M., Arif, M.S., Shahzad, S.M., Hussain, S., Akhtar, M.J., Abbas, F., 2015. Spatial distribution of pollutant emissions from crop residue burning in the Punjab and Sindh provinces of Pakistan: uncertainties and challenges. Environ. Sci. Pollut. 22 (21), 16475−16491.

Ishaque, H., 2017. Is it wise to compromise renewable energy future for the sake of expediency? an analysis of Pakistan's long-term electricity generation pathways. Energy Strategy Rev. 17, 6−18.

Jan, I., Ullah, S., Akram, W., Khan, N.P., Asim, S.M., Mahmood, Z., Ahmad, M.N., Ahmad, S.S., 2017. Adoption of improved cookstoves in Pakistan: a logit analysis. Biomass Bioenergy 103, 55−62.

Kanter, D.R., Chodos, O., Nordland, O., Rutigliano, M., Winiwarter, W., 2020. Gaps and opportunities in nitrogen pollution policies around the world. Nat. Sustain. 3, 956−963.

Kummu, M., de Moel, H., Porkka, M., Siebert, S., Varis, O., Ward, P.J., 2012. Lost food, wasted resources: global food supply chain losses and their impacts on freshwater, cropland, and fertiliser use. Sci. Total Environ. 438, 477−489.

Lassaletta, L., Billen, G., Grizzetti, B., Garnier, J., Leach, A.M., Galloway, J.N., 2014. Food and feed trade as a driver in the global nitrogen cycle: 50-year trends. Biogeochemistry 118 (1−3), 225−241.

Liu, C., Wang, Q., Zou, C., Hayashi, Y., Yasunari, T., 2015. Recent trends in nitrogen flows with urbanization in the Shanghai megacity and the effects on the water environment. Environ. Sci. Pollut. Res. 22, 3431−3440.

Mauzerall, D.L., Sultan, B., Kim, N., Bradford, D.F., 2005. NOx emissions from large point sources: variability in ozone production, resulting health damages and economic costs. Atmos. Environ. 39 (16), 2851−2866.

Mosier, A.R., Syers, J.K., Freney, J.R., 2004. Nitrogen fertilizer: an essential component of increased food, feed, and fiber production. In: Agriculture and the Nitrogen Cycle: Assessing the Impacts of Fertilizer Use on Food Production and the Environment. Island Press, pp. 3−15.

National Electric Power Regulatory Authority, 2020. State of Industry Report 2020. Government of Pakistan, pp. 7−23.

Nayyar, Z.A., Zaigham, N.A., Qadeer, A., 2014. Assessment of present conventional and nonconventional energy scenario of Pakistan. Renew. Sustain. Energy Rev. 31, 543−553.

Niazi, T., 2003. Land tenure, land use, and land degradation: a case for sustainable development in Pakistan. J. Environ. Dev. 12 (3), 275−294.

Norton, J., Ouyang, Y., 2019. Controls and adaptive management of nitrification in agricultural soils. Front. Microbiol. 10, 1931.

Pakistan Bureau of Statistics, 2018. Population Census. Statistics Division, Government of Pakistan. http://www.pbs.gov.pk/pco-kpk-tables. (Accessed 15 December 2020).

Phillips, S.M., 2012. Dietary protein requirements and adaptive advantages in athletes. Br. J. Nutr. 108 (2), 158–167.

Pimentel, D., Pimentel, M., 2003. Sustainability of meat-based and plant-based diets and the environment. Am. J. Clin. Nutr. 78, 660–663.

Poore, J., Nemecek, T., 2018. Reducing food's environmental impacts through producers and consumers. Science 360 (6392), 987–992.

Rahman, A.K.M., Zhang, D., 2018. Effects of fertilizer broadcasting on the excessive use of inorganic fertilizers and environmental sustainability. Sustain. Times 10, 759.

Raney, T., Gerosa, S., Khwaja, Y., Skoet, J., Steinfeld, H., McLeod, A., Opio, C., Cluff, M., 2009. The State of Food and Agriculture: Livestock in the Balance. Food and Agriculture Organization of the United Nations, Rome, Italy.

Raza, S., Zhou, J., Aziz, T., Afzal, M.R., Ahmed, M., Javaid, S., Chen, Z., 2018. Piling up reactive nitrogen and declining nitrogen use efficiency in Pakistan: a challenge not challenged (1961–2013). Environ. Res. Lett. 13 (3), 034012.

Rehman, S.A., Cai, Y., Siyal, Z.A., Mirjat, N.H., Fazal, R., Kashif, S.U.R., 2020. Cleaner and sustainable energy production in Pakistan: lessons learnt from the Pak-TIMES model. Energies 13 (1), 108.

Rehman, A., Deyuan, Z., 2018. Pakistan's energy scenario: a forecast of commercial energy consumption and supply from different sources through 2030. Energy Sustain. Soc. 8, 26.

Saeed, A., Abbas, M., Manzoor, F., Ali, Z., 2015. Assessment of fine particulate matter and gaseous emissions in urban and rural kitchens using different fuels. J. Anim. Plant. Sci. 25, 687–692.

Shah, M.H., Shaheen, N., Nazir, R., 2012. Assessment of the trace elements level in urban atmospheric particulate matter and source apportionment in Islamabad, Pakistan. Atmos. Pollut. Res. 3 (1), 39–45.

Sharma, L.K., Bali, S.K., 2017. A review of methods to improve nitrogen use efficiency in agriculture. Sustain. Times 10, 51.

Sheikh, A.M., 2003. Pakistan Energy Yearbook: Hydrocarbon Development Institute of Pakistan. Ministry of Petroleum and Natural Resources, Government of Pakistan, Islamabad.

Shirazi, S.A., 2012. Temporal analysis of land use and land cover changes in Lahore-Pakistan. Pakistan Vis. 13 (1), 187.

Smil, V., 2000. Feeding the World: A Challenge for the Twenty-First Century. MIT Press, USA.

Smil, V., 2002. N and food production: proteins for humans' diets. Ambio 31, 126–131.

Steinfeld, H., Gerber, P., Wassenaar, T., Castel, V., Rosales, M., De Haan, C., 2006. Livestock's Long Shadow: Environmental Issues and Options. Food and Agriculture Organization of the United Nations (FAO), Rome, Italy.

Stewart, W.M., Dibb, D.W., Johnston, A.E., Smyth, T.J., 2005. The contribution of commercial fertilizer nutrients to food production. Agron. J. 97 (1), 1–6.

Tahir, S.N.A., Rafique, M., Alaamer, A.S., 2010. Biomass fuel burning and its implications: deforestation and greenhouse gases emissions in Pakistan. Environ. Pollut. 158 (7), 2490–2495.

Ti, C., Xia, L., Chang, S.X., Yan, X., 2019. Potential for mitigating global agricultural ammonia emission: a meta-analysis. Environ. Pollut. 245, 141–148.

Tilman, D., Clark, M., 2014. Global diets link environmental sustainability and human health. Nature 515, 518−522.

Townsend, A.R., Howarth, R.W., Bazzaz, F.A., Booth, M.S., Cleveland, C.C., Collinge, S.K., et al., 2003. Human health effects of a changing global nitrogen cycle. Front. Ecol. Environ. 1 (5), 240−246.

Vitousek, P.M., Naylor, R., Crews, T., David, M.B., Drinkwater, L.E., Holland, E., et al., 2009. Nutrient imbalances in agricultural development. Science 324 (5934), 1519−1520.

World Health Organization, 2007. Protein and Amino Acid Requirements in Human Nutrition World Health Organization (WHO) Technical Report Series Number 935.

Yang, X., Geng, J., Li, C., Zhang, M., Tian, X., 2016. Cumulative release characteristics of controlled-release nitrogen and potassium fertilizers and their effects on soil fertility, and cotton growth. Sci. Rep. 6, 39−30.

Zhang, W.F., Dou, Z.X., He, P., Ju, X.T., Powlson, D., Chadwick, D., et al., 2013. New technologies reduce greenhouse gas emissions from nitrogenous fertilizer in China. Proc. Natl. Acad. Sci. U.S.A. 110 (21), 8375−8380.

# Trends in nitrogen use and development in Pakistan

5

**Abdul Wakeel[1], Aysha Kiran[2], Muhammad Rizwan Shahid[1], Zunaira Bano[2], Munir Hussain Zia[3]**

[1]*Institute of Soil and Environmental Sciences, University of Agriculture Faisalabad, Punjab, Pakistan;* [2]*Department of Botany, University of Agriculture Faisalabad, Punjab, Pakistan;* [3]*R&D Department, Fauji Fertilizer Company Limited, Rawalpindi, Punjab, Pakistan*

## 5.1 Chemical fertilizer offtake in Pakistan

Pakistan is an agricultural country and its economy is dependent on agricultural productions. Although livestock, dairy, and other allied animal sectors have great contribution to Pakistani agriculture, however, the role of crop production is very vital. Due to semiarid climate and low soil organic matter, crop production is dependent on synthetic chemical fertilizers. As per the national data collected by National Fertilizer Development Corporation, total fertilizer offtake has been presented considering nutrients nitrogen (N), phosphorus (P), and potassium (K), and their changing trends along with population increase have been elaborated in the following sections.

### 5.1.1 Fertilizer offtake by nutrients

Like many other developing countries, use of synthetic fertilizers was started in 1960s and from then the use of synthetic fertilizer is increasing with some drop downs. During 2016−17, maximum total nutrient offtake has been observed which was ∼5 million tons. However, it surprisingly dropped to 3.5 million tons in 2018−19. The possible reasons might be high fertilizer prices as compared to previous year with about 10% losses in crop production keeping the farmers restricted to low use of fertilizers due to financial constraints (economic survey of Pakistan, 2019−20). This strong correlation also indicates the high dependency of crop production on synthetic fertilizer. Being an agricultural country, Pakistan is contributing greatly to global food security. Pakistan is among top 10 users of synthetic fertilizers; nevertheless, fertilizer applied per unit land area is far less than many countries and Pakistan is placed at 58th number in global ranking. However, low per unit fertilizer application does not mean high fertilizer use efficiency because average per unit yield is also low compared to most of the agricultural countries. Even in India with similar land type, climate and agricultural practices is getting more per unit

Nitrogen Assessment. https://doi.org/10.1016/B978-0-12-824417-3.00002-2

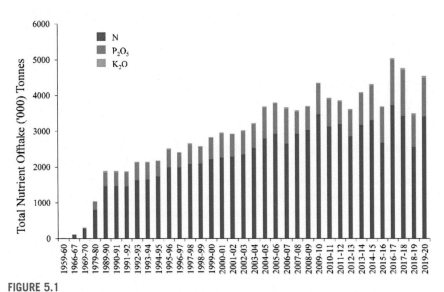

**FIGURE 5.1**

Chronological fertilizer offtake by nutrients in Pakistan (NFDC, 2020).

yield than that in Pakistan. Another important point is imbalance use of fertilizers (Fig. 5.1).

Crops demand for K is similar to that of N and even more for many crops, but the use of potassic fertilizers is negligible if compared with N fertilizers where more than 45% soils are deficient in plant available K. This can be one of the factors responsible for low fertilizer use efficiency in crop production (Shahzad et al., 2019). Although there is a steep chronological increase in fertilizer use, the increment in average per unit yield is stagnant since many years. Relatively high use of N fertilizers is due to lack of awareness among farmers, subsidized rates of urea, and comparatively high price of other fertilizers. Another important factor for low N use efficiency (NUE) can be associated with very generalized N fertilizer recommendation without considering the soil capability for crop production.

### 5.1.2 Crop-wise nutrient use

Use of fertilizer varies with crop species, production potential of cultivars, and economic value of the crops. It is not similar for all crops considering per unit and ultimately total use of fertilizer at national level. Considering national crop production of Pakistan, fertilizers use can be divided into six groups. Five major crops, i.e., wheat, rice, cotton, maize, and sugarcane, utilize 93% of the total fertilizer used in the country. Remaining 7% of fertilizer is used for other crops including various vegetables and fruits. Wheat is the highest consumer of fertilizers, i.e., ∼2.3 million tons (Fig. 5.2A), because it is a main staple crop and grown on >9 million hectares of land which is 40% of total cultivated area of Pakistan. However, more than 70% is

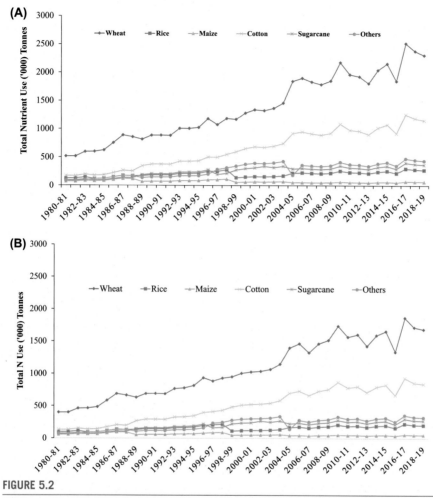

**FIGURE 5.2**

(A) Chronological nutrient (as N + P₂O₅+K₂O) use by five major crops cultivated in Pakistan. Major crops include wheat, rice, cotton, maize, and sugarcane and others include potato, tobacco, fruits, and some more vegetables (NFDC, 2020). (B). Chronological nitrogenous fertilizer use by five major crops cultivated in Pakistan. Major crops include wheat, rice, cotton, maize, and sugarcane and others include potato, tobacco, fruits and some more vegetables. The N is calculated through the ratio of all three nutrients N, P, and K (NFDC, 2020).

N fertilizer in the form of urea, about 25% is phosphatic fertilizer, and negligible amount of potassic fertilizers. The second highest user of synthetic fertilizer is cotton, using 1.1 million tons of fertilizers with 73% nitrogenous fertilizers. In cotton-growing area, excessive use of nitrogenous fertilizers has been reported and greater atmospheric losses due to high temperature (>45°C) in these areas has been observed (Shahzad et al., 2019). Rice and maize require 0.3 and 0.07 million tons fertilizer per annum, respectively, with high percentage of nitrogenous fertilizers use. The trends in Fig. 5.2A and B revealed that there is a chronological increase in fertilizer for all six crop groups; however, there is a steep increase for wheat and cotton using major portion of total N fertilizer used in the country with rapid further increase with high rate. It is direly needed to enhance the NUE to fulfill the food requirements. Stagnant yields instead of increasing trend of fertilizer application also indicate the low response to fertilizer application possibly due to declining soil health and imbalanced fertilizer use. The ascending trend of fertilizer use is not due to enhanced acreage of these crops because there is a negligible change in land use and acreage but due to descending fertilizer use efficiency. Ultimately, it is also decreasing farm productivity and creating unrest for farmers and environmental unsustainability due to release of reactive N species in water and atmosphere.

## 5.2 Demographic projections—2070

Globally it has been estimated that population is increasing at a steepest rate and projected to increase almost in the same manner until 2070 as indicated by the orange line for estimated and blue lines for predicted population growth (Fig. 5.3). A total increase of 7.4 billion population from 1950 to 2015 has been observed which is projected to increase up to 11.2 billion by the year 2070.

### 5.2.1 Population projection in comparison to south Asia

The data from eight south Asian countries which includes India, Bangladesh, Pakistan, Afghanistan, Bhutan, Sri Lanka, Nepal, and Maldives regarding population estimates were compared in context of trends in population growth during next 50 years (Fig. 5.3). Pakistan is the only country which has the steepest curve both about estimated and predicted values as compared to all other countries except Afghanistan. Afghanistan and Bhutan as compared to Pakistan have a varied trend because in Afghanistan during the year 1986, population decreased from 13 to 11 M and in Bhutan during years 1992−1996 a fall of 0.51M from 0.53 M was observed, as shown by orange line in the graphs. Afterward population in these countries raised up to 33.7 M in Afghanistan and 0.78 M in Bhutan by the year 2015 and projected to increase up to 70.4 and 0.86 M prior to 2070, respectively.

Nepal and Maldives show a less steep curve (orange lines), whereas for predicted increase in population more explicit curves can be observed mentioned declining trend. Sri Lanka shows bit contrasting trend ranging from estimated (orange line) to projected values (blue lines), it shows sharper curve compared with Pakistan and other countries. Contrary to other countries, Sri Lanka has also population decline within next 20 years. The most populous countries of south Asia are India, Pakistan, and Bangladesh with current population of 1383, 221, and 165 million

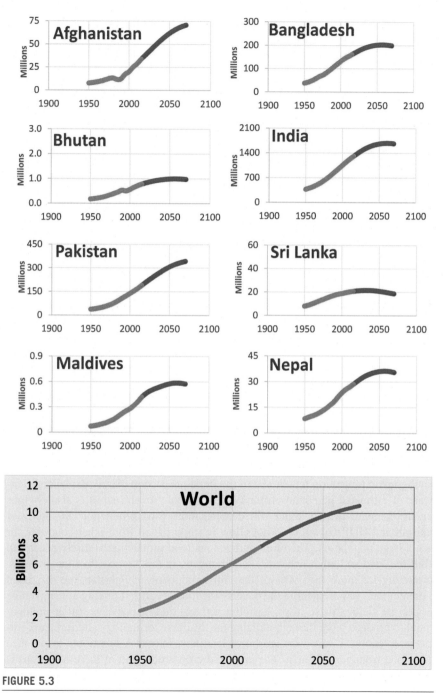

**FIGURE 5.3**

Comparison of estimated (orange line 1950–2015) and predicted (*blue line* 2016–70).

people, respectively. The population growth rate is 0.99%, 2.0%, and 1.01% for India, Pakistan, and Bangladesh which will add up 282, 122, and 34 million more people till 2070 to the population of these countries, respectively. The lowest values have been observed for Maldives and Bhutan which are 0.41 and 0.78 M, respectively, whereas Nepal, Afghanistan, and Sri Lanka show a moderate trend.

### 5.2.2 Food production considering population trends

Pakistan has a significant number of total population and the highest growth rate in south Asia as discussed above. It shows that after next 50 years, more food and shelter will be required for about $\sim 122$ million more people with limited land resources. It can only be possible by utilizing the available resources in efficient way because there is no expansion capacity and neither it is recommended considering possible environmental and ecosystem concerns.

A major part of agricultural commodities produced in Pakistan are used locally as a food which include wheat, cotton, rice, maize, and sugarcane. Also, the vegetables and fruits are majorly used to fulfill the local requirement. It means even if the export remains the same, there will be a need to produce more food, especially, in association with population projection, from limited available resources. Another best option may be to reduce the wastage of food or to utilize the food in economical way to reduce the pressure of more production on available resources. About 53% fruits and vegetables are wasted, whereas 24% cereals are waste after harvest. By conserving wasted food, pressure on further food production with enhanced N fertilizer use may be reduced.

## 5.3 Trends in N fertilizer use considering food production trends

Demographic projection mentioned in the previous section clearly explains the increase of estimated population from 1980, so is the growth rate. It is also obvious that the increase in population will demand more food production. Economy of Pakistan and GDP growth does not allow to import food from other countries for such a huge population. Therefore, indigenous food production is direly needed to fulfill the food and other necessary requirements. In this section, five major crops, accounting for more than 90% of the crop production in Pakistan, are discussed to predict their requirement in the country for food security. Looking at the average per hectare yield of these crops, it is observed that these crops have still enough potential to produce more through utilizing the available technology and further scientific intervention. Limited expansion of land area may also contribute to produce more food from per unit area. The predicted crop production of these five crops is purely based on needs of the population during next 50 years, and it is further assumed that most of the required food will be produced from local resources.

### 5.3.1 **Wheat**

The trend in nitrogen use is elevating with the passage of time to increase yield of wheat in response to fulfill the demand of increasing population (Fig. 5.4). Green bars show the wheat production in million tons based on population projections and red line shows the N use in 1000 tons. Data from 1980 to 2019 are observed data and from 2020 to 2070 it was predicted to anticipate the future demand of food and nitrogen use for wheat production. During the years 1986, 2005, 2006, 2009, 2012, 2013, 2015, 2016, and 2017 a sharp ups and downs in grain yield can be observed and can be correlated with variation in N fertilizer use. Based on predicted production requirement to feed the predicted population, it has been observed that the increase in population will also affect significantly the N use; however, this increase in N use will also increase the N losses, if the NUE could not be enhanced. Nitrogen use prediction shows that N use in wheat to produce the required wheat grains will be doubled by 2070 which means that $\sim$2 million tons of extra N in the field. Considering 40% losses can cause a huge impact on environment through nitrogenous gases emission.

### 5.3.2 **Rice**

Rice is the second most important cereal and staple crop for many countries globally and in some areas of Pakistan. NUE in rice is about 30%−40%. Fig. 5.5 presents the trend of N use correlated with rice production from 1980 to 2017 in million tons filled in green columns, whereas red line indicates the use of N in 1000 tons. The

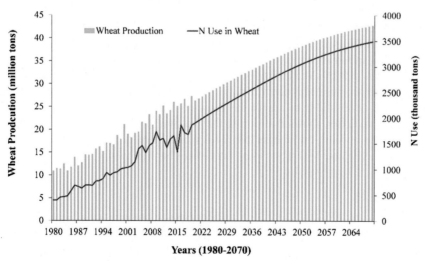

**FIGURE 5.4**

Predicted nitrogen (N) use in accordance with the increase in wheat production from 1980 to 2070 (NFDC, 2020).

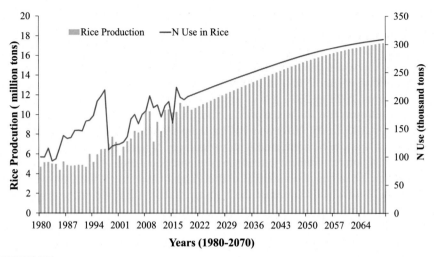

**FIGURE 5.5**

Predicted nitrogen (N) use in accordance with the increase in rice production from 1980 to 2070 (NFDC, 2020).

data of rice production from 2020 to 2070 are predicted data based on population growth considering association of population and rice production from 1980 to 2019. The trend of NUE in rice is different from all other four major crops which can be observed from already mentioned figure where during 1997 the use of N was >219 thousand tons and then in the next year it fell to ~113 thousand tons, whereas the production of rice in these 2 years did not show much change. The following years, use of N remained constant around 120 thousand tons to 126 thousand tons till the year 2002. Afterward a sharp increase up to ~175 thousand till 2005 with a transient drop in 2006 (158.96 thousand tons) and then touched 200 thousand tons in 2009. Maximum N was used in 2016, when it crossed 220 thousand tons. The variable use of N fertilizers is dependent on the fertilizer and rice prices which were never stable to attract the farmers' attention. The population trend–based rice prediction shows that during 2070 > 300 thousand tons N will be used and taking in account the current NUE for rice a great N losses are expected in the form of $NH_3$ and $N_2O$ if the things could not be improved.

### 5.3.3 Maize

Significant amount of fertilizers is required for maize production to fulfill the local needs and Fig. 5.6 depicts the extended N use correlated with predicted maize production. The bars show the production of crop in million tons and the line shows the N use in thousand ton. The data from 1980 to 2019 is calculated and from 2020 to 2070 is calculated using regression model based on predicted maize production to feed the increasing population using extended population graph till 2070. The usage

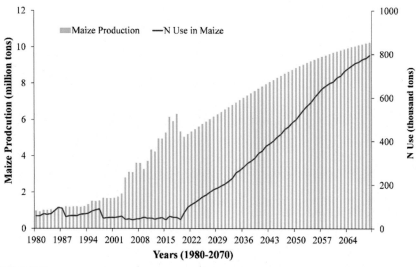

**FIGURE 5.6**

Predicted nitrogen (N) use in accordance with the increase in maize production from 1980 to 2070 (NFDC, 2020). Data available between 2000 and 2020 seem unrealistic; however, it is the only quotable data available in Pakistan.

of N with respect to the maize production is high in earlier year from 1980 to 1996 which starts decreasing from 1997 and remained constant till 2019. This is purely based on available information; however, it is not understandable why use of fertilizer was not increased when the production has been increased manifold. Nevertheless, these data have been used just to show off a prediction of N use for future scenario but these prediction for predicted N use for maize may be considered as suspected until the more authenticated information of fertilizer use in past two decades. The highest N use has been observed in the year 1986 which was 96.41 thousand tons and lowest values were observed in the year 2006 which was 39.74 thousand tons. The decrease in N use did not impose much effect on the production of maize. From the year 2004–18 maize productions increased from 1.09 million tons to 6.31 million tons. The trend in maize N use was quite uncertain during the past years. But in predicted data it is observed that there is sharp increase in N use with respect to maize production.

### 5.3.4 Sugarcane

The use of N fertilizer has been correlated with the sugarcane production using regression model (Fig. 5.7). Red line shows the trend of N use in 1000 tons per year and green bars shows the sugarcane production in million tons per year. The data from the year 1980–2019 are estimated, whereas the data from the year 2020–70 are predicted using regression model. In the above-mentioned graph, in

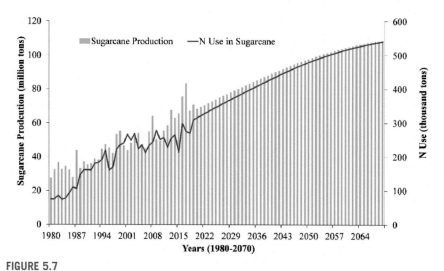

**FIGURE 5.7**

Predicted nitrogen (N) use in accordance with the increase in sugarcane production from 1980 to 2070 (NFDC, 2020).

relevance to N fertilizer use an undulating trend is observed like in different years 1996, 2006, and 2015 and then there was a steep drop in the amount of N use as compared to the previous years. In 1996, values decreased from 219.05 thousand tons to 161.23 thousand tons and a much greater drop from 265.84 to 211.95 thousand tons can be observed during the period of 6 years, i.e., from 2001 to 2006. Highest use of 298.40 thousand tons N fertilizers was observed in the years 2016 which dropped to 269.32 thousand tons in 2018. The trend for production of sugarcane is undulating with some sharp high and low levels of transitions, nevertheless with an increasing N use trend. Highest production of 83.33 million tons was observed during the year 2017 and lowest sugarcane production, i.e., 27.86 million tons in 1986. Most of the sugar produced in Pakistan is indigenously used, therefore considering population growth induced national sugar consumption predicts a huge increase in N use to sustain the increasing population (Fig. 5.7).

## 5.3.5 Cotton

Cotton is among the key cash crops of Pakistan, especially in Punjab and Sindh provinces. The data from 1980 to 2019 are estimated, whereas from 2020 to 2070 it is predicted to anticipate the future demand and supply scenario from sustainability point of view. Red line shows the trend in N use in 1000 tons and green bars show the cotton production in million tons (Fig. 5.8). A smoother trend can be seen in nitrogen use during the period of 20 years (1980−2000). During this period the nitrogen use has increased from 133.28 1000 tons to 522.81 1000 tons. Then in the following years from 2004 to 2019 sharp transitions can be observed. A

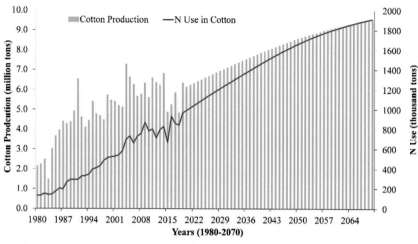

**FIGURE 5.8**

Predicted nitrogen (N) use in accordance with the increase in cotton production from 1980 to 2070 (NFDC, 2020).

concurrent cotton production trend can be seen with highest recorded production 7.28 million tons in 2004, whereas lowest amount 1.48 million tons in 1983. Considering future scenario, it is important to use N in crop production in an efficient way to fulfill the demand of the people.

### 5.3.6 Total N use

Time series data on the N use in accordance with the total production of five major crops (wheat, rice, cotton, maize, and sugarcane) from 1980 to 2070 have been illustrated in Fig. 5.9. The data from 1980 to 2019 are estimated, whereas from 2020 to 2070 are predicted to anticipate the future demand and N use in crops for sustainable production. Red line shows the trend in nitrogen use in 1000 tons and green bars show the total production of crops in million tons.

The data showed that growing population has significant impact on the crop production leading to enhanced crop production for food security. To fulfill the food demand, the excessive use of N fertilizers in crops will be increased in future and the excessive use of N fertilizers in crops will significantly impact the environment negatively. The release of greenhouse gases especially nitrous oxide ($N_2O$) from high N fertilizer use contribute to climate change. There is need for action to tackle these changes, and for sustainable production the NUE for all these crops can be enhanced by identifying and developing N-efficient cultivars. Using smart N fertilizers with high efficiency including inhibitors, nanofertilizers, etc., will also contribute to sustainable crop production with less Nr emissions. Use of N fixing crops can be included in the detailed strategies to minimize the losses of N.

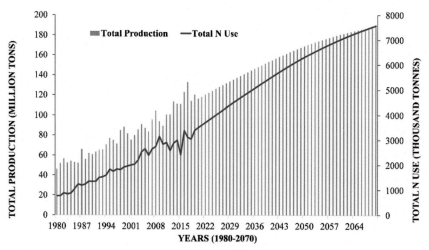

**FIGURE 5.9**

Predicted total nitrogen (N) use in accordance with the increase in production of major crops to fulfill the demand 1980–2070 (NFDC, 2020).

## 5.4 Trends in N fertilizer use considering agricultural interventions

Organic matter in soils is considered a major reserve of nitrogen, phosphorus, sulfur, and micronutrients. However, cultivated soils in most irrigated areas of Pakistan are deficient in organic matter, which necessitates the use of N fertilizers. This is because of hot and dry weather during most part of the year; and frequent ploughing that promote decomposition of organic matter. Compared to arid regions, soil organic matter can easily be maintained/enhanced with less organic residue in cold/temperate as well as wet/moist regions.

Since most of the organic matter values fall well below 2%, therefore, to draw country scale map, we lowered down the threshold around 1% value to see how many soils are extremely deficient in organic matter. The map shows that soil organic matter is generally lower than 1%, except for certain pockets of the rice belt of eastern Pakistan and Sindh province, and the maize and potato belt in Okara and Sahiwal districts (Fig. 5.10). The spatial variation reveals that most of the extremely deficient soils are in Punjab followed by Sindh. The map could also help plan strategies for improving farm practices across specific regions to improve N fertilizer advice.

The detail of major interventions that boosted agriculture growth and therefore N fertilizer consumption across the country has been explained in the following section.

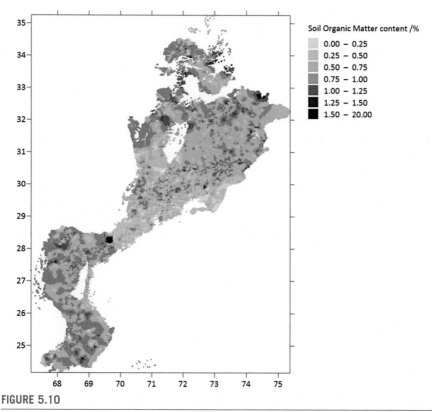

**FIGURE 5.10**

Kriging prediction of organic matter (proxy of N) in topsoil of Pakistan.

*Data courtesy: Fauji Fertilizer Company Ltd., Pakistan.*

## 5.4.1 Green revolution

During initial decade of the Green Revolution, wheat production in Pakistan registered an increase of 84%, i.e., from 3.7 million tone (1959−60) to 6.8 million ton (1968−69). Thus, using dwarf type of wheat introduced in Pakistan in 1960s, the country was able to outperform its neighboring countries. Alongside, for the next 15 years, there was an increase in irrigation supply through tube wells, and canals in addition to fertilizers.

Other than the benefits in terms of higher yields, the dwarf varieties had a weakness, because of the DELLA proteins, that they were unable to absorb N as traditional varieties could, so they required higher amount of fertilizer to grow (Li et al., 2018). Because of higher per acre net income (156%−222%), low price of fertilizer input, and high response in terms of increased yields, fertilizer N use steadily grew over the next two decades. It was negligible at start of 1960s, i.e., 0.03 million tons however, N use touched level of 1.1 million ton in 1985−86; 2 million ton in 1995−96; 3 million ton in 2008−09; and touched a peak of 3.7 million ton in

2016–17 (Fig. 5.1). Consumption of N remains almost same for both the cropping seasons, i.e., Rabi and Kharif. Provincial share in N consumption is dominated by Punjab, followed by Sindh, KP, and Balochistan. Higher N use (per unit area) in Sindh compared to Punjab can be attributed to the irrigated area, which is almost 100%.

### 5.4.2 Development of Mangla and Tarbela water reserves

The rate of adoption of high yielding varieties took a great momentum during the period 1966–76. During this period, the proportion of area sown to new varieties increased from meager 1% to more than 80% with a peak observed in 1988. Because of the development of Mangla and Tarbela dams, supply of irrigation water in the Punjab increased markedly between the decade 1967–76 and the decade that followed. For example, irrigation water supply to wheat increased from 46 acre inch in 1967 to 63 acre inches in 1986 (Byerlee and Siddiq, 1994). This improved irrigation supply created matching demand for fertilizers (Figs. 5.11 and 5.12). For example, during this period, fertilizer application improved from less than 10 kg nutrients per hectare in 1966 to 130 kg nutrients per hectare in 1986 in irrigated wheat (Byerlee and Siddiq, 1994). The growth in fertilizer consumption @ 10% per annum for the two decades could not be expected in case of short supply of irrigation water. Of the total nutrients consumed, N fertilizers share was around 60%–70% during the period of 1976–86, whereas it was 90% during the 1st decade, i.e., 1967–76. Large-scale comparative data survey for wheat reported in literature also suggest

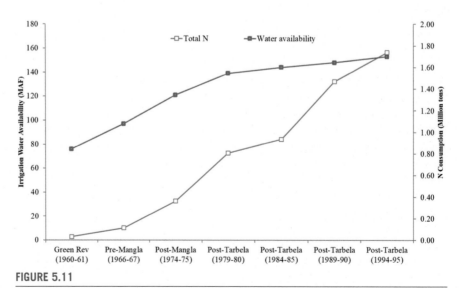

**FIGURE 5.11**

Irrigation water availability versus N fertilizer consumption (Pre-Mangla and Post-Tarbela).

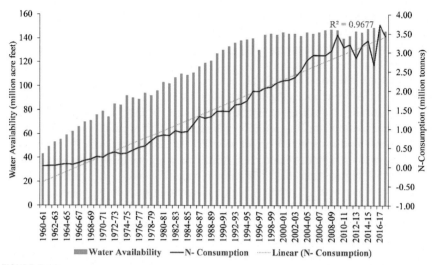

**FIGURE 5.12**

Historical water availability in Pakistan in relation to overall N fertilizers consumption.

*Data source: WAPDA; Agric Stat Pak; and NFDC Fertilizer Reviews.*

N consumption shifts by almost 100%, i.e., from 48 kg ha$^{-1}$ (year 1976–77) to 93 kg ha$^{-1}$ (year 1987–88).

The impact of the development of these water reserves can even be realized in present era since the reserves ensure smooth supply of irrigation water during *rabi* season to wheat crop which consumes 50% fertilizers of the country.

### 5.4.3 Adoption to Bt cotton and other crop hybrids and overuse of N

Beyond increasing acreage of high N-fertilizer consuming crops like sugarcane, maize, cotton, etc., introduction of Bt cotton in Pakistan and similarly other hybrids (e.g., maize, rice, and tomato) further boosted N fertilizer consumption in the country. This is because of higher N requirement of such crops compared to the traditional cultivars. For example, in Punjab, recommendation of N fertilizer to Bt cotton falls in the range of 70–100 kg acre$^{-1}$ (average 85 kg acre$^{-1}$) which is 34% higher than that of N dose recommended for non-Bt cotton, i.e., 63.5 kg per acre N. For 2.8 million hectare area across the country, additional N consumption (compared to non-Bt) stands at 149,000 ton, assuming that all the cotton being grown in Pakistan is Bt. Contrary to recommendations set by Agric. Extension Punjab (N equivalent to four bags of urea), farmers in Punjab apply six bags of urea and/ or even more as a common practice to encourage new boll formation and thus better harvest of Bt cotton. Similarly, in Sindh province, farmers generally apply 269 kg N ha$^{-1}$ to Bt cotton compared to Sindh Agriculture Extension Department recommendations (192 kg ha$^{-1}$). Similar to this, hybrid maize is also getting popularity beyond its core area of Okara and Sahiwal and is currently being grown over 30% area (approx. 0.3 million ha) dedicated to maize crop in the country. On an average, hybrid maize requires 37 kg per hectare more N compared to synthetic maize (199 kg N vs. 235 kg N ha$^{-1}$).

In case of hybrid rice, 83 kg N per acre has been recommended by research scientists from UAF (but not by Agric. Extension Deptt., Punjab) which is 20% higher than recommended N for Irri type rice (N 69 kg acre$^{-1}$), and 50% higher than that of the recommendations for Basmati (55 kg N acre$^{-1}$). This would translate into additional N consumption especially in Sindh where 90% of hybrid rice of the country is grown. In case of hybrid tomato, N fertilizer recommendation is 90 kg per acre, which is 55% higher than that of N recommended for traditional cultivars of this crop (58 kg N per acre). Extremely high consumption of N in tunnel crops across the country is yet unaccounted for simply because of the lack of data.

Keeping above in view, it can safely be inferred that adoption of hybrid/improved cultivars of Bt cotton, hybrid rice, maize, and tomato has boosted consumption of N fertilizer in the country. The additional use of N through hybrid crops is further augmented by overuse of N fertilizers in the country that could partially be linked to lack of farmers' knowledge about balanced fertilization, see Table 5.1.

### 5.4.4 Subsidy on N fertilizers

Cost of production for various crops in Pakistan is relatively high in comparison to other countries. For example, during the last 5 years (2015–20), fertilizer expense alone accounted for 20%–25% in wheat; 8%–13% in sugarcane; 12%–17% in cotton; 13%–17% in coarse rice; and 12%–14% in fine rice against total cost of production. Fertilizer sector was regulated and heavily subsidized by mid-1980s, after which privatization and deregularization initiatives were implemented. Regression analysis demonstrates that yields of wheat, rice, cotton, maize, and sugarcane are positively affected by fertilizer subsidy in Pakistan (Danish et al., 2017).

Subsidy on N fertilizers was initiated in the late 1950s to generate farmers' interest and throughout 1960s it expanded, significantly. During 1975–77, subsidies on local fertilizers were increased from Rs. 10 million to Rs. 140 million (Chaudhry et al., 1995). From 1984 to 85 to 1992–93, local fertilizers subsidies were reduced from Rs. 830 million to 110 million through exclusion of N fertilizers under Structural Adjustment Program. Later, subsidies on phosphatic and potassic fertilizers were also abolished in 1995 and 1997, respectively. Subsidies were not disbursed by GoP on local fertilizers from 1993 to 2005, and on imported fertilizers from 1996 to 2003. Similar trend was observed after 2005 when GoP announced Rs. 50 billion for subsidy to farmers on fertilizers in 2011–12, highest ever during the period of 1980–81 to 2011–12 (Danish et al., 2017). The increasing trend could not sustain after 2011 and more than 60% decline in subsidies was observed over the next 5 years.

The balance of subsidies had always been tilted toward N rather than equal weightage for other nutrients, especially P, K, and Zn. This is the reason that N

**Table 5.1** Survey data about overuse of N in certain districts of Sindh.

| Crop | Urea (bags) | Avg. urea dose (per acre) | Total N from urea (kg acre$^{-1}$) | CAN (bags) | Total N from CAN (kg acre$^{-1}$) | NP (bags) | Total N from NP (kg acre$^{-1}$) | Total N from urea, CAN, and NP (kg acre$^{-1}$) | Recommended N by the sindh Agric. Deptt (kg acre$^{-1}$) | Estimated overuse (%) | District(s) | Remarks |
|---|---|---|---|---|---|---|---|---|---|---|---|---|
| Bt cotton | 3–4 | 3.5 | 80.5 | 1 | 13 | 1 | 11 | 105 | 78 | 34 | All Sindh | General practice |
| Sugarcane | 4–5 (grown as sole crop) | 4.5 | 103.5 | 1 | 13 | 1 | 11 | 128 | 110 | 16 | Matiari, Tando Muhammad Khan, Tando Allahyar, and Sanghar | Intercropping of sugarcane in onion is largely followed in Matiari, Tando Allahyar, and part of T.M. Khan district |
| | 3–4 (inter-cropped in onion) | 3.5 | 80.5 | 1 | 13 | 1 | 11 | 105 | 87 | 20 | Hyderabad, Matiari, Tando Muhammad Khan, Tando Allahyar, and Sanghar | |
| Onion | 3–4 | 3.5 | 80.5 | 2 | 26 | 1 | 11 | 118 | 87 | 35 | Matiari, Hyderabad, Tando Allahyar, and Tando Muhammad Khan | Some farmers adjust dose of urea following a myth. If FYM is used, urea is applied @ 4–5 bags per acre. On the other hand if FYM is not used, urea is applied @ 6–8 bags per acre |
| Chillies | 2–3 | 2.5 | 57.5 | 4 | 52 | 3 | 33 | 143 | 110 | 30 | Mirpur khas, Umer kot including Jhudo, Kunri | Picking mostly starts in August. Farmers apply urea or NP after every picking. Mostly use NP and CAN |

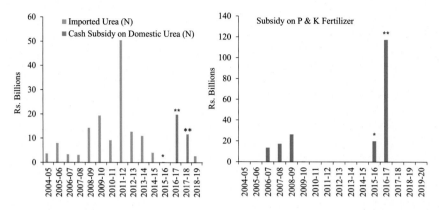

**FIGURE 5.13**

Fertilizer subsidy on locally manufactured and imported fertilizers (NPK) in Pakistan (NFDC, 2020), *Budget allocation for phosphate and potash, **Data regarding phosphate and potash subsidy for 2016—17 and 2017—18 show subsidy claims submitted by fertilizer companies and not actual disbursement. Notes: a) The figures for the fiscal years 2006—07, 2007—08, 2008—09, 2009—10, and 2016—17 also include subsidy given on P and K fertilizers which is 22.9, 28.2, 34.2, 0.6, and 19.64 billion Rs., respectively, b) The figures do not include subsidy given by Govt of Punjab against phosphatic fertilizers (DAP Rs. 500, NP Rs. 200, and SSP Rs. 200 per 50 kg bag); potassic fertilizers (SOP Rs. 800, MOP Rs. 500, and NPKs Rs. 300 per 50 kg bag); fertilizers through coupon scheme for the period 2018—19. For the last 2 years (2017/18 and 2018/19), Rs. 1335 million have been disbursed by the Govt of Punjab against DAP; 92 million against NP, SSP, and NPK; and Rs. 183 million against SOP/MOP fertilizers. Moreover, an amount of Rs. 1400 million has been allocated for the year 2020 (per comm with DG Agric Ext, Punjab), c) The figures do noy include subsidy given on feedstock gas to fertilizer factories which amount to Rs. 334.28 billion for the period 1995—2014.

fertilizers consumption took huge leap over time in the country, whereas even today farmers are using less P and K in relation to N. For GoP to discourage excessive use of N fertilizer, the right opportunity to introduce subsidy on phosphatic fertilizers was in 1999 when Fauji Fertilizer Bin Qasim Ltd established first ever phosphatic fertilizer production unit in the country.

Over the last 44 years (1975—2018) about Rs. 296.2 billion has been offered as subsidy for locally manufactured as well as imported fertilizers (Fig. 5.13). Out of this, Rs. 105.54 billion was allocated to P and K fertilizers; therefore, exclusive subsidy on N fertilizers stands at Rs. 190.66 billion.

## 5.5 Future N fertilizer research trends in Pakistan

Spending on agriculture research in Pakistan is low in comparison to other developing countries of Asia. Pakistan invests 0.25%—0.29% of its agricultural GDP on research compared to 0.35% in Bangladesh, 0.4% in India, and 0.6% in China.

Opposite to this, in the developed world, investment on R&D is typically 2%—3% of its agricultural GDP, e.g., 2.5% in Japan (Flaherty et al., 2013).

Uptake and utilization of the applied N fertilizer generally varies between 30% and 60% depending upon soil texture, irrigation frequency, climatic conditions, crop type, and the physicochemical and biological reactions within soil environment. Generally, among cereals it is the lowest for rice because of flooding conditions and high temperatures during summer. Fertilizer use efficiency is lowest in Pakistan for cotton, wheat, and rice considering partial factor of productivity (Shahzad et al., 2019). Among the recommended measures to improve NUE, following five measures are being followed globally in addition to many others:

**a.** Enhanced efficiency fertilizers (EEFs).
**b.** Genetics and management practices assuring maximum economic yields.
**c.** Precision agriculture technologies to sense crop needs and improve application.
**d.** Increased use of on-farm measures evaluating nutrient use efficiency.
**e.** Decision support tools applying science at the farm level.

In addition to development of breeding traits like roots morphology and surface area, internal utilization within plant, and Best Farm Management Practices like split application, band placement, adoption of soil test—based site-specific fertilizer recommendations, scientists in collaboration with the fertilizer industry are putting focus on development and testing of EEFs which can be classified into three broad categories:

**a.** Slow/Controlled Release Fertilizers
**b.** Urease (chemical additives), and nitrification inhibitors (NIs) (biological and chemical inhibitors)
**c.** Other specialty fertilizers (highly water-soluble fertilizers, special grades like NPKs, etc.)

The environmental risks associated with the use of fertilizers include leaching as nitrates; greenhouse gas emissions (nitrous oxides and ammonia); and nitrogen (plus phosphorus) losses as run off into water bodies. There are two key aspects focused while developing EEFs to address such risks. One is to develop a physical barrier, viz. polymer material, so that nutrient release could be controlled over a few days up to few months. The barrier thus coated over the fertilizer may reduce nutrient element concentration depending upon the thickness. Another approach is to combine some chemical with the fertilizer to affect its uptake by the crop. Several nitrification and urease inhibitors have been identified that may influence biological process through which nutrients are made available to crop, i.e., potential role of microorganisms within soil environment (Ashraf et al., 2019).

### 5.5.1 Slow or Controlled Release Fertilizers

This category of EEF has been divided into two groups by Trenkel (1997). One group is comprised of encapsulated or coated fertilizers, such as polymer-coated urea (PCU) or sulfur-coated urea (SCU). The other group is formed by condensation products of urea and urea aldehydes, for example, crotonylidene diurea (CDU), isobutylidene diurea (IBDU), and urea formaldehyde (UF).

Although most of the coatings have been developed for soluble N fertilizers, especially urea (prilled and granular forms), however, other fertilizers like NPKs have also been reported to be coated for slow/controlled release features. Sulfur in its elemental form (molten form at 150°C) was introduced by Tennessee Valley Authority probably as the first coating material for urea in late 1960s. However, development of the cracks within coating layer significantly degraded its efficiency (Shaviv, 2005). This was addressed through introduction of wax/other polymer materials layer(s) as a sealant.

At present, mostly thin polymer layers are coated at commercial scale in developed world but not in Pakistan. The polymer layer plays the role of controlling the permeability of irrigation water to fertilizer N granule and therefore release urea-N via diffusion process, slowly through the swelling polymer membrane (unlike SCU which releases urea through pinholes) over time. Thickness of the coating decides whether the nutrient N would be released over a specified time period but this also reduces elemental N concentration in the fertilizer prill/granule (Irfan et al., 2018). This ultimately helps enhance NUE and therefore losses of N to the environment are reduced, significantly.

There are contradictory reports about benefits of such improved products in terms of yield increase. Nonetheless, these EEFs are of high interest to rice growers in Pakistan, as they would reduce labor costs involved in split N application besides agronomic and environmental benefits.

The other group under EEF are PCU formaldehyde—based nitrogenous products. These could be either water insoluble or water soluble or a mixture of both. CDU, IBDU, and urea formaldehyde (UF, nitrogen 38%) are common examples of this group, although latter is the most popular in this category. The products' solubility and release of N depends upon ratio with which the paraformaldehyde(s) have been reacted with standard urea. Several soil factors like texture, moisture, temperature, pH, microbes, etc., play a role in mineralization of the nitrogen (to ammonium form) from such polymer-based urea formaldehydes. Some other fertilizer products applied for specialized action in plants under particular soil conditions might also be considered under this category, for example, urea super granules and zinc-fortified urea, etc. Engro and Fauji Fertilizer company (FFC) have tested efficiency of urea supergranules. However, it could not market the same due to social aspects of its adoption/application in paddy fields.

### 5.5.2 Stabilized fertilizers (nitrification and urease inhibitors)

There are several organic, biological, and inorganic substances that can either affect microbial activity or the chain reactions happening during nitrification process, prolonging nitrogen availability to crop(s). Most common categories include urease inhibitors and NIs. Urease inhibitors reduce activity of certain bacteria that release urease enzyme, and this helps in slow conversion of urea-N into ammonium form. Similarly, NIs affect the activities of nitrosomonas and nitrobacter so that nitrogen transformation (oxidation) from ammonium form into nitrite, and then nitrate forms could be slowed down. As mentioned earlier that impact of these inhibitors on grain yield is also dependent upon several other environmental factors, for example, soil temperature, moisture, texture, organic carbon contents, etc. According to Mahmood et al. (2017), higher temperatures under calcareous conditions like that of Pakistan, 4-Amino-1,2,4-triazole (ATC) is more effective compared to other commercial NIs viz. nitrapyrin, 3,4-dimethylpyrazol phosphate (DMPP), and dicyandiamide (DCD). Similar efficiency of ATC has also been reported by Ali et al. (2012), who tested eight NIs, i.e., 1H-benzotriazole; 4-amino-1,2,4-triazole; benzothiazole; 3-methylpyrazole-1-carboxamide; 4-bromo-3-methylpyrazole; pyrazole; lignosulfonic acid, molecular weight 52,000, 6% S; and lingo-sulfonic acid, molecular weight 12,000, 2% S. The best strategy to evaluate the effectiveness of slow/controlled released fertilizers and NIs is employment of $N^{15}$ isotopic techniques This technique may clearly distinguish whether the nitrogen taken up by test crop has originated from improved fertilizer test batch or from soil pool compared to control treatment, where standard N fertilizer is used. This would also quantify the exact uptake in various parts of the crop. Ashraf et al. (2004) employed $N^{15}$ isotope to study NUE from legumes, nonlegumes, and inorganic N sources in flooded rice.

Extensive research was carried out in the 1960s and 1970s to identify and test effectiveness of numerous compounds for their possible role as "nitrification inhibitors." The most popular examples reported in the literature include 3,4-dimethylpyrazole phosphate (DMPP); dicyandiamide (DCD, $H_4C_2N_4$); and 2-chloro-6-(trichloromethyl) pyridine (Nitrapyrin). Another type of NI is calcium carbide−sourced acetylene. To enhance NUE of urea-based N products, use of urease inhibitors has also been commercialized that reduce $NH_3$ volatilization from urea hydrolysis. Many micronutrients like Mn, Cu, and Zn and boric acid are also able to inhibit urease activity within soil environment. Most of the urease inhibitors can prevent urea hydrolysis for 1 or 2 weeks, during which time the urea fertilizer should ideally be incorporated into the soil with water (rainfall or irrigation) or mechanical methods (Chien et al., 2009). For better agronomic performance, many researchers have stressed on combined use of nitrification and urease inhibitors.

### 5.5.3 Local market—enhanced efficiency N fertilizers

Global Market of EEF including specialty fertilizers is projected to reach $20 billion by the year 2020. Locally in Pakistan, only three commercial scale fertilizer products (NUREA—SCU 36%N; zincated urea; and neem-oil−coated urea (NCU)) fall under the category of EEFs. Out of the three products, only zincated urea and NCU are sparsely available in the market, whereas SCU is no longer available. FFC's R&D Department and Engro Fertilizers have signed research agreements with local and foreign universities and research Institutes to test and develop EEF products. Even though field trial results for NCU product are inconsistent in India,

Indian Government has made it compulsory for every manufacturer to coat all its urea produce with neem oil and has also regulated its bag weight at 45 kg instead of 50 kg that was used in past for standard urea. Such strategies can be equally workable in Pakistan as well. Nevertheless, careful validation of neem oil–coated fertilizers is required to observe its impact on Nr emissions in different crop rotations.

### 5.5.4 Crop rotations and nitrogen management

Crop rotation is the agronomic practice of growing crops on the same piece of land in sequence. It has great significance while considering the efficient N management. Among its key benefits nutrient cycling, improved soil properties, and weed control are important (Asseng et al., 2014). Furthermore, its impact on N mineralization and reduced use of chemical fertilizer has direct impact on reactive nitrogen (Nr) emissions to atmosphere and nitrate leaching. The drastic increase in use of chemical nitrogenous fertilizer has not only affect the soil health but also contribute to atmospheric pollution, therefore use of leguminous crops in crop rotation is key to reduce N fertilizer application without affecting the crop production and environmental sustainability. Inclusion of legumes in crop rotation has revealed increase in yield and reduced emissions in many studies. Along corn, use of soybean not only increased the yield up to 20% for corn and 7% for soybean but also reduced 35% $N_2O$ emission (Behnke et al., 2018). Legumes inclusion in crop rotation leads to reduce carbon emissions and other greenhouse gases as well and is very use for changing climate scenario (Hazra et al., 2020). In Pakistan, diversification in crop rotation is not well planned, so is the use of legumes in crop rotation and as cover crops, although a variety of legumes are available to be cultivated with excellent marketing potential. Legumes as cover crops provide a substantial amount of N to succeeding crop lowering the dependency on chemical fertilizers promoting the agricultural and environmental sustainability.

### 5.5.5 Nano fertilizers for enhanced N use efficiency

Synthesis and practical implications of nanomaterials is the great innovation of recent times. It has beneficial use in different fields of life including agriculture. Nanotechnology has come up as a powerful developmental tool for new technological product. Nanofertilizers are a promising technology for agriculture and have been tested in soil as well as foliar application. Use of nanotechnology for fertilizer can be categorized into three parts: nanosized fertilizers itself, use of nanosized carriers of nutrients to enhance their efficiency, and nutrients encapsulated with nanoparticles films to reduce the release (Usman et al., 2020). The results of nanofertilizers revealed beneficial outcomes when tried for various crop species (Mejias et al., 2021). Generally granular nitrogenous fertilizers are used which has greater losses of reactive nitrogen (Nr) to atmosphere and use of nanomaterials in N fertilizers can lead to better NUE lowering the Nr emissions to atmosphere as well as nitrate leaching. Although some research trials are being conducted in

Pakistan for the use of nanofertilizers, however, there is great potential and scope for it to explore the best nanofertilizer strategies for enhanced NUE.

### 5.5.6 Economics of enhanced efficiency N fertilizers

Globally, fertilizer industry is making huge efforts in R&D to bring down cost of coated fertilizer. For example, the environmentally smart nitrogen (ESN) being marketed in United States is about 20%–40% higher (US$ 0.17–0.44 premium per kg product) in price compared to standard urea. The farm economics also depends upon ratio with which such product(s) are mixed with that of standard urea, for example, 50%–80% ESN is being recommended for mix with 50%–20% standard uncoated urea in the United States. Locally in Pakistan, SCU (N 38%, $K_2O$ 2%, S 15%) marketed in past by Safi Chemicals under brand name of "NUREA" could not gain momentum due to its high price (Rs. 2150 for 25 kg bag—Dec 2018 price). For NCU, a local company has charged a premium of Rs. 20 per bag, even though its production cost is higher than that of the premium. Another local company also test marketed NCU in 45 kg bags at Rs. 100 less compared to standard urea. Together, both the companies marketed about 30,000 tons of the product during the years 2016 and 2017/18. Another local company (JBL) is also marketing a different category of bio-fortified fertilizers under brand name of "Nutraful," i.e., biologically enhanced fertilizers that could probably be labeled as specialty fertilizers within broader domain of EEF. Market price of Nutraful urea is about Rs. 300–400 per 50 kg bag—higher than that of standard urea. Urea super granules are another fertilizer applied at 80% of recommended N dose showing similar performance in paddy fields (along with additional yield benefit) compared to full dose of commercial urea.

### 5.6 Conclusion

Population of Pakistan is growing rapidly with growth rate of about ∼2% which is, perhaps, higher than all south Asian countries, inducing a pull for more food production. Most of the food items in Pakistan are dependent on five major crops, i.e., wheat, rice, maize, sugarcane, and cotton. Due to limited land resources, the only option to enhance food production is increase in per hectare yield which has great potential; however, it will demand more fertilizer use due to poor soil health especially soils with low organic matter. Nitrogenous fertilizers contribute more 65% to total applied fertilizers with very low use efficiency leading to further N losses as reactive nitrogen (Nr) species. There is great need to enhance the N fertilizer use efficiency keeping in view the increasing demand of food production to feed the huge population growing rapidly. Subsidy on efficient and value-added N fertilizers is needed to promote smart fertilizer with low Nr emissions will reduce the Nr emissions from agricultural productions.

# References

Ali, R., Kanwal, H., Iqbal, Z., Yaqub, M., Khan, J.A., Mahmood, T., 2012. Evaluation of some nitrification inhibitors at different temperatures under laboratory conditions. Soil Environ. 31 (2), 134–145.

Ashraf, M., Mahmood, T., Azam, F., Qureshi, R.M., 2004. Comparative effects of applying leguminous and non-leguminous green manures and inorganic N on biomass yield and nitrogen uptake in flooded rice (*Oryza sativa* L.). Biol. Fertil. Soils 40, 147–152.

Ashraf, M.N., Aziz, T., Maqsood, M.A., Bilal, H.M., Raza, S., Zia, M., Mustafa, A., Xu, M., Wang, Y., 2019. Evaluating organic materials coating on urea as potential nitrification inhibitors for enhanced nitrogen recovery and growth of maize (*Zea mays* L.). Int. J. Agric. Biol. 22, 1102–1108.

Assent, S., Zhu, Y., Basso, B., Wilson, T., Cammarano, D., 2014. Simulation modeling: applications in cropping systems. In: Encyclopedia of Agriculture and Food Systems, pp. 102–112.

Behnke, G.D., Zuber, S.M., Pittelkow, C.M., Nafziger, E.D., Villamil, M.B., 2018. Long-term crop rotation and tillage effects on soil greenhouse gas emissions and crop production in Illinois, USA. Agric. Ecosyst. Environ. 261, 62–70.

Byerlee, D., Siddiq, A., 1994. Has the green revolution been sustained? The quantitative impact of the seed-fertilizer revolution in Pakistan revisited. World Dev. 22 (9), 1345–1361.

Chaudhry, M.G., Sahibzada, S.A., Salam, A., 1995. Agricultural input subsidies in Pakistan: nature and impact. Pakistan Dev. Rev. 34 (4), 711–722.

Chien, S.H., Prochnow, L.I., Cantarella, H., 2009. Recent developments of fertilizer production and use to increase nutrient efficiency and minimize environmental impacts. Adv. Agron. 102, 261–316.

Danish, M.H., Tahir, M.A., Azeem, H.S.M., 2017. Impact of Agriculture Subsidies on Productivity of Major Crops in Pakistan and India: A Case Study of Fertilizer Subsidy. Punjab Economic Research Institute (PERI), Planning and Development Department, Government of the Punjab. Publication No. 427.

Flaherty, K., Stads, G.J., Srinivasacharyulu, A., 2013. Benchmarking Agricultural Research Indicators across Asia-Pacific. ASTI Synthesis Report of International Food Policy Research Institute, Washington, D.C.

Hazra, K.K., Nath, C.P., Ghosh, P.K., Swain, D.K., 2020. Inclusion of legumes in rice–wheat cropping system for enhancing carbon sequestration. In: Carbon Management in Tropical and Sub-tropical Terrestrial Systems. Springer, Singapore, pp. 23–36.

Irfan, M., Niazi, M.B.K., Hussain, A., Farooq, W., Zia, M.H., 2018. Synthesis and characterization of zinc-coated urea fertilizer. J. Plant Nutr. 41, 1625–1635.

Li, S., Tian, Y., Wu, K., Ye, Y., Yu, J., Zhang, J., Liu, Q., Hu, M., Li, H., Tong, Y., Harberd, N.P., Fu, X., 2018. Modulating plant growth–metabolism coordination for sustainable agriculture. Nature 560, 595–600.

Mahmood, T., Ali, R., Lodhi, A., Sajid, M., 2017. 4-Amino-1,2,4-triazole can be more effective than commercial nitrification inhibitors at high soil temperatures. Soil Res. 55 (7), 715–722.

Mejías, J.H., Salazar, F.J., Pérez, L., Hube, S., Rodriguez, M., Alfaro, M.A., 2021. Nanofertilizers: a cutting-edge approach to increase nitrogen use efficiency in grasslands. Front. Environ. Sci. 9, 52.

NFDC, 2020. Statistics National Fertilizer Development Centre. http://www.nfdc.gov.pk/stat. html. Accessed June 2020.

Shahzad, A.N., Qureshi, M.K., Wakeel, A., Misselbrook, T., 2019. Crop production in Pakistan and low nitrogen use efficiencies. Nat. Sustain. 2, 1106–1114.

Shaviv, A., 2005. Controlled release fertilizers. In: IFA International Workshop on Enhanced-Efficiency Fertilizers, p. 2830. Frankfurt, Germany.

Trenkel, M.E., 1997. Controlled-Release and Stabilized Fertilizers in Agriculture. (FAO) IFA, Paris.

Usman, M., Farooq, M., Wakeel, A., Nawaz, A., Cheema, S.A., ur Rehman, H., Ashraf, I., Sanaullah, M., 2020. Nanotechnology in agriculture: current status, challenges and future opportunities. Sci. Total Environ. 721, 137778.

## Further reading

Association of American Plant Food Control Officials (AAPFCO), 1995. Official Publication No. 48. Association of American Plant Food Control Officials, Inc., West Lafayette IN.

Bilal, B., Niazi, M.B.K., Jahan, Z., Hussain, A., Zia, M., Taqi, M., 2020. Coating materials for slow release of nitrogen from urea fertilizer: a review. J. Plant Nutr. 43 (10), 1510–1533.

Blaylock, A., 2010. Enhanced Efficiency Fertilizers. Colorado State University Soil Fertility Lecture, 10-20-2010. Agrium Advanced Technologies, Loveland, CO.

Hall, A., 2005. Benefits of enhanced-efficiency fertilizer for the environment. In: IFA International Workshop on Enhanced-Efficiency Fertilizers, pp. 28–30. Frankfurt, Germany.

# Nitrogen emissions from agriculture sector in Pakistan: context, pathways, impacts and future projections

**Muhammad Irfan[1], Nighat Hasnain[2]**

[1]*Soil and Environmental Sciences Division, Nuclear Institute of Agriculture (NIA), Tandojam, Sindh, Pakistan;* [2]*Agrilenz Ltd., London, United Kingdom*

Other than macroeconomic data on fertilizer/manure use trends in Pakistan, there is limited context-specific data available on N usage or the environmental and health impacts of excessive N in the environment. Reasons for this are the lack of understanding among policy makers, practitioners, and researchers of why excess N in the environment is a problem: how transformations of excess N take place in the soil—plant system and how these transformations can lead negative impacts on the environment, health, agricultural productivity, and farm profitability. The focus in this chapter therefore is as much about understanding these issues around N as it is to understand the trends in N usage within the country and potential impact. This chapter will identify:

- Why N in the environment is a global concern and the context for Pakistan.
- The fate of N through the transformations that take place in soil—plant system and the processes that can lead to losses from the system.
- The impact of this N on air quality, ecosystems, biodiversity, food production capacity, and human health at a global level. The focus in this chapter is mainly on gaseous emissions of $NH_3$ and $NO_x$.
- The historical changes in N use and emission trends from agriculture in Pakistan.
- The gaps in our knowledge around N usage, its impact within our food production and consumption systems.
- Key areas where strategies need to be developed to improve our understanding of the agricultural systems in Pakistan, the N pathways, the field level practices, the extent of the N problem, and the identification of good practices which can inform future strategies.

Due to limited data availability for Pakistan, global and regional datasets have been used to demonstrate the enormity of anthropogenic perturbation on the N cycle and resulting implications for the environment and public health. These issues are likely to exist in Pakistan, however, there is still little attention being paid to

Nitrogen Assessment. https://doi.org/10.1016/B978-0-12-824417-3.00008-3

them. Strategies need to be developed and implemented that reduce demand for N or those that limit the emissions from entering the bio-geochemical system. In this chapter, we have highlighted the areas that require urgent attention in order to develop strategies to reduce N in the environment from agriculture.

## 6.1 The N issue and the Pakistan context

### 6.1.1 The N issue

Nitrogen is a crucial nutrient that increases productivity in agriculture. Since the invention of Haber–Bosch process, which converts the relatively inactive N in the atmosphere to a reactive form, $N_r$, global consumption of $N_r$ in agriculture has increased substantially and is projected to increase further in the future. Global estimates indicate that use of reactive compounds of N will be 150% higher than in 2010, with the agriculture sector accounting for 60% of this increase (Martínez-Dalmau et al., 2021).

High synthetic $N_r$ inputs when applied in excess to crop demand lead to $N_r$ losses, a lower N utilization efficiency with a negative impact on production (Liu et al., 2010; Shahzad et al., 2019). This is compounded by a poor management and utilization of livestock manure, leading to further losses of $N_r$ in the system. The amount of $N_r$ released into the biosphere through anthropogenic means is prodigious and is estimated at 120 Tg $yr^{-1}$ which is twice that of the N fixed by all natural terrestrial processes, i.e., 63 Tg $yr^{-1}$ (Fowler et al., 2013). In the early 21st century, anthropogenic perturbation of the global N cycle has contributed about two-thirds of the annual flux of $N_r$ into the atmosphere (Fowler et al., 2015) and with a growing population, South Asia has become a hotspot for N emissions (Tian et al., 2014). The past century has also experienced rising demand of livestock products, which has accelerated the demand for fodder crops, hence fertilizer use resulted in the production of more animal waste (Dangal et al., 2017). Increase in application of high N fertilizer to food and feed crops coupled with poor or no management of livestock manure is increasing N emissions from agriculture in South Asia.

$N_r$ losses in the system result in costly impacts on the environment and human health (Raza et al., 2018; Shahzad et al., 2019; Sutton et al., 2021). Sutton et al. (2013) estimated an annual damage of $\sim 800$ billion \$US globally due to N pollution. In addition to low economic returns, nitrate leaching or runoff to water resources and N losses as $NH_3$ and $NO_x$ to the atmosphere is leading to serious degradation of air quality, water quality, loss of biodiversity, and an increase in health risks (Liu et al., 2020).

### 6.1.2 Pakistan context

Pakistan is the sixth most populous country in the world and second in the South Asia. Production of food has increased significantly over the decades spurred on by fiscal incentives as well as the growth in the local fertilizer manufacturing industry, in particular urea production. Despite significant increase in crop and livestock production, a recent report on food security by the International Food Policy Research Institute ranked Pakistan 106 out of 119 developing countries, having about 20% of the country population undernourished (von Grebmer et al., 2018). With a population over 218 million and the threat of malnutrition increasing, it follows that crop and livestock production will need to increase to meet food requirements of the growing population.

Most arable soils of Pakistan are considered to be N deficient with low bioavailability of N to plants. Low rainfall together with low organic matter content in soils (typically between 0.5% and 1.0%) are among the major factors behind this N limitation. As the role of $N_r$ application in maintaining and increasing crop productivity is pivotal (Irfan et al., 2018), regular addition of synthetic fertilizers are often prescribed to compensate for this deficiency. Consequently, the application of $N_r$ has increased over the years, leading to surplus N in soil and its subsequent loss to the environment. Financial subsidies have played a significant role here as they helped distort market prices, making fertilizers relatively cheap to use rather than risk a loss in production. The net result is the indiscriminate usage of N fertilizers as well as the subsequent losses of $N_r$ in the system.

These losses are not well documented and it so is difficult to attribute the losses to sectors or regions of the country. However, using macroeconomic data on fertilizer usage at a national level, Raza et al. (2018) calculated the corresponding impact on air emissions between 1961 and 2013. They estimated a significant increase in $NH_3$, $N_2O$, and NO emissions from 70, 10, and 1 Gg N $yr^{-1}$ in 1961 to 1023, 155, and 46 Gg N $yr^{-1}$ in 2013, respectively. Although it is difficult to allocate it by agricultural sector or spatially, this is a significant rise and is likely to have an impact on the environment, health, and the sustainability of the agricultural productivity.

The following section discusses the fate of $N_r$ in the environment, the transformations, potential for losses, and the impact these losses have on the environment, human health, and agricultural productivity. Due to the limited data for Pakistan, the sections provide useful insights on the $N_r$ issue at a general level. They will also help the reader understand why national-level data though a good starting point are not adequate to understand the extent or the nature of the $N_r$ problem in the country. These still need to be understood within the Pakistan context where land ownership and management, holding size, livestock ownership, and management patterns can vary significantly from those found in regions of the world where this problem is recognized better and there are more data available. There is even less information on low-input systems such as rangelands where low N input land management by transhumant communities plays a significant part in the livestock economy. The section on information gaps and key needs discusses these further and should be used to help develop options that improve $N_r$ utilization, reduce losses to the environment, and improve food security.

## 6.2 Nitrogen transformations in soil and gaseous losses

### 6.2.1 Nitrogen transformations

The N cycle is one of the most important nutrient cycles found in the terrestrial ecosystem and understanding the N pathways, its transformations and utilization is imperative to manage $N_r$ in the environment (Fig. 6.1). Molecular N ($N_2$) is highly unreactive and hard to extract from the atmosphere. Plants cannot access this unless converted to more reactive forms such as $NH_4^+$ or $NO_3^-$ via lightning followed by deposition or through biological fixation by N-fixing bacteria, algae and fungi (Fowler et al., 2013; Ghaly and Ramakrishnan, 2015). Reactive N circulates from the atmosphere to the terrestrial and aquatic biosphere into organic compounds and then back into the atmosphere. The N containing organic matter added to the soil is transformed into soluble inorganic form ($NO_3-N$ and $NH_4-N$) by the process termed mineralization. The opposite process is immobilization in which inorganic N is converted into proteins by microorganisms and plants. Plants can assimilate $N_r$ ($NO_3^-$ and $NH_4^+$) resulting from natural mineral deposits, fertilizers, manures, decomposition of organic matter, and atmospheric deposition (Rashid and Memon, 2010).

Within the soil, a major portion of soil N (>98%) is in an organic immobile form which plants cannot utilize directly. This organic N undergoes the processes of mineralization to make it bioavailable to the plants, the end product of mineralization is $NH_4^+$. The ammonium (added as synthetic fertilizer or mineralized from

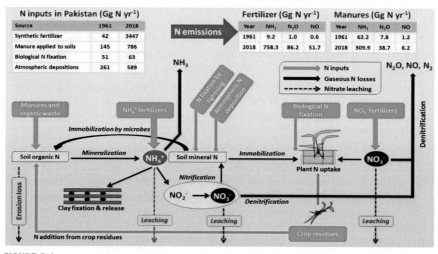

**FIGURE 6.1**

Nitrogen transformations in soil—plant system indicating different N sources, processes involved (mineralization, immobilization, nitrification, and denitrification) and N losses. Numerical values show comparison for various N inputs and emission losses from fertilizer and manures in Pakistan during the years 1961 and 2018.

organic N) may undergo nitrification to form nitrates under aerobic conditions. These nitrates can be taken up by plants/microbes but excess can also be lost to the atmosphere, for example, in anaerobic conditions, $NO_3^-$ ions are converted to gaseous forms such as $N_2$ through a series of reduction reactions called denitrification. In addition, microbial mediated denitrification can also reduce $NO_3$ to produce molecular N. Incomplete denitrification also generates $N_2O$ and NO which are released in the air. The $N_r$ losses from agro-ecosystems originate through deprotonation of $NH_4^+$ to $NH_3$ (Marschner, 2012).

Some of the negative charges in $NO_3^-$ ions cannot be adsorbed by the negatively charged soil colloids and hence move downward freely with drainage water and readily leached from the soil. The loss of N via $NO_3^-$ leaching may cause soil acidification and coleaching of cations like $Ca^{2+}$, $Mg^{2+}$, and $K^+$. However, in arid and semiarid climates, $NO_3^-$ leaching seldom occurs except after heavy irrigation (Brady and Weil, 2012).

In soil, $N_2O$ is produced both via nitrification under aerobic conditions and denitrification under anaerobic conditions. In nitrification process, bacterial oxidation of ammonium ($NH_4^+$) takes place to nitrate ($NO_3^-$). During nitrification, $NH_4-N$ is converted to hydroxylamine ($NH_2OH$) followed by NOH and then $NO_2$, while $N_2O$ is generated at both steps of $NH_2OH$ and NO. In first step, $NH_4^+$ is converted into nitrite ($NO_2^-$) by *Nitrosomonas*, while in second step $NO_2^-$ is converted into $NO_3^-$ by *Nitrobacter* (Leininger et al., 2006). On the other hand, denitrification transforms $NO_3^-$/ NO to $N_2$ or $N_2O$. Denitrification is usually mediated by aerobic heterotrophic bacteria having the ability to reduce $NO_3^-$ to gaseous forms such as $N_2O$.

These cooccurring processes of mineralization, immobilization, nitrification, and denitrification in the soil determine the nutrient availability to plants (Marschner, 2012). The soil processes are in turn governed by soil pH, temperature, moisture, N application rate, and several microbial-mediated nitrification and denitrification reactions (Chen et al., 2015). For the system to operate with optimal plant uptake and low N losses, these processes need to be in equilibrium. However, as soil processes are not that well understood nor are management of manure and fertilizer application perfectly aligned with weather, soil conditions, and plant needs, there is a propensity to apply more N to overcompensate this uncertainty. This causes an excess $N_r$ in the system which is eventually lost to the ecosystem through the pathways shown in Fig. 6.1.

### 6.2.2 N transformations and interaction with C cycle and soil food web

Nutrient cycling is fundamental to ecological functioning and is expected to strongly impact food web stability. It can provide over 50% of the total nutrient supply of the food web leading to a strong enrichment effect thereby promoting species persistence in nutrient poor ecosystems and may lead to a paradox of enrichment at high nutrient inputs (Quevreux et al., 2018). Soil biological processes greatly depend on soil organic carbon (C). A strong interaction between C and N dynamics is usually expected as N transformations in soil are mainly driven by microbial activity, while organic C acts as a source of energy (Abrar et al., 2020; Ashraf et al., 2020). The content of organic C is influenced by quality and quantity of the organic material giving rise to different rates of N transformations within the soil profile. Addition of organic material or crop residues with wide C:N ratio favors microbial immobilization of N ($NO_3^-$ and $NH_4^+$). Higher immobilization may reduce $N_2O$ emissions or reduce the total N losses (Vanlauwe et al., 2002). On decomposition of organic materials with narrow C:N ratio, rapid release of $NO_3^-$ and $NH_4^+$ may occur and can increase the production of $N_2O$. Soil characteristics and land use can affect soil food webs and their corresponding C and N fluxes. Nitrogen addition in agroecosystems can dramatically disrupt soil microbial communities that regulate nutrient cycling processes. Manipulation of soil food web structure under N addition in turn could alter microbial functional activities and process rates (Crowther et al., 2015).

Anthropogenic $N_r$ to the terrestrial biosphere has increased up to five-fold over the past century. Increasing $N_r$ greatly affects C and N cycling by lowering pH (Raza et al., 2021) and increasing N availability which can directly influence soil food web and soil processes (Bobbink et al., 2010). The N enrichment—induced changes can have both negative and positive impacts on C and N cycling. Increase in N availability favors vegetative growth and the production of root exudates and labile C (Chen et al., 2016). This, in turn, may increase the C supply for soil microbes changing the C:N ratio and impacting mineralization rates by affecting soil microorganisms, soil acidification, and nutrient contents. Soil acidification can limit soil microbial activity as well as C and N cycling by changing the level of soil base cations (such as $Ca^{2+}$ and $Mg^{2+}$) and $H^+$ ions (Rousk et al., 2010).

### 6.2.3 N losses from agriculture

Ammonia is a key pollutant in agricultural systems and accounts for around 50% of global $NH_3$ emissions. It is produced through direct volatilization from surface application of N fertilizers like urea and manure, as well as complex physical, chemical, and biological soil processes such a hydrolysis of urea to produce ammonium carbonate followed by $NH_3$ which rapidly escapes into the atmosphere (Cameron et al., 2013). The rate of $NH_3$ volatilization is primarily influenced by soil and environmental conditions with more rapid increases in high pH soils and at high temperature (Brady and Weil, 2012), although addition of N to acidic and neutral pH soils can also release $NH_3$. Warmer climatic conditions and high pH of soils in Pakistan are therefore likely to increase the volatilization rates. The application process for N can also influence $NH_3$ volatilization rates with high values for broadcasting operations and low values for injection with incorporation into soil, especially if incorporated within 24 hours (Maqsood et al., 2016).

Another gaseous pollutant is nitrous oxide, $N_2O$, a potent and long-lived greenhouse gas with a Global Warming Potential of 265 to that of $CO_2$. Besides global warming, it can react with oxygen in the stratosphere to produce $NO_x$ and nitric acid followed by a series of reactions that deplete stratospheric ozone concentrations. Agricultural practices are the major source of anthropogenic $N_2O$ emissions. Increasing use of N fertilizer and livestock manure is contributing significantly to $N_2O$ emissions from cropping systems (Sarabia et al., 2020).

### 6.2.3.1 Nitrogen emissions from cropping sector

The major contributor of global N emissions to environment is agriculture sector, and similarly in Pakistan, about 65% of N emissions ($NH_3$ or $N_2O$) are being released by agriculture sector (www.fao.org. accessed 17 July 2021). In cropping systems, agricultural activities such as fertilizer use, livestock manure spreading, and burning crop residues are major contributors of gaseous $N_r$ emissions usually as $NH_3$ and $NO_x$. Ammonia is mainly released from activities such as fertilizer use and livestock manure management, whereas fossil fuel combustion in agricultural machinery, vehicles, and industrial sources can be important sources of $NO_x$ (Huang et al., 2014). $NO_x$ is also emitted from nonfossil fuel or natural sources such as lightning, biomass burning, and microbial processes in soil (Lu et al., 2019). In addition, agricultural operations such as tillage, planting, harvesting, handling, and transportation of crops release small amounts of $NO_x$ to the air. Factors that significantly accelerate $NH_3$ emissions are high soil N content, soil pH, and temperature (Riddick et al., 2016). Additionally, biomass burning, though highly episodic in nature, accounts for more than 10% of the global $NH_3$ emissions (Barnes and Rudzinski, 2006).

### 6.2.3.2 Nitrogen emissions from livestock sector

The generation and use of manure in both cropping and livestock systems are an important source of $N_r$ in agro-ecosystems. Amount and chemical compositions of manure generated by different animals on different farm types are regarded as $N_r$ hot spots (Steinfeld et al., 2006). $N_r$ emissions take place at all stages of manure management from production, storage, and application to soil (Rotz, 2004). If animals are housed, the amount and type of $N_r$ loss differs greatly with the type of animal housing. For instance, $NO_x$ emissions may occur through both nitrification and denitrification processes when animals are housed on unpaved surfaces (Rotz et al., 2014).

During storage of manure, again $NH_3$ is released depending on the level of mineralization of manure and the design of storage facility. Livestock manures are generally solid or liquid slurry. Solid manure can be stored in stacks where considerable N losses as $NH_3$ may occur depending upon the stack dimensions and storage time. Over time it may develop a surface crust which reduces $NH_3$ loss but may favor nitrification and denitrification processes leading to higher $NO_x$ emissions (Rotz and Leytem, 2015). Storing liquid manure or slurry (more mineralized) in lagoons under anaerobic conditions can increase total N loss; however, denitrification processes can also lead to almost 50% N lost as $N_2$. Storing slurry in a tank may reduce $NH_3$ loss depending upon the exposed surface area.

Manure is generally spread through broadcasting or spraying across the soil surface. This leads to significant quantities of $NH_4-N$ remaining in the manure which rapidly volatilizes as $NH_3$. Factors controlling the amount of $NH_3$ produced and emission rate include manure temperature, pH, and air speed across the surface. Incorporation of manure reduces $NH_3$ emissions but can lead to substantial $NO_3^-$ losses depending on soil conditions and method of incorporation (Rotz and Leytem, 2015).

In summary, $N_r$ emissions from agro-ecosystems are mostly dependent upon agronomic and soil management practices as well as environmental conditions. Substantial N losses occur in the form of $NH_3$ emissions, when fertilizers and manures are applied on soil surface as well as from livestock manure management systems. Timing and method of application of fertilizer, soil temperature, wind conditions, and pH impact these emissions (Rotz et al., 2014). Consequently, these need to be addressed in both the arable and the livestock sector to achieve reductions in $NH_3$ emissions.

---

## 6.3 Impacts of elevated N emissions ($NH_3$, $NO_x$)

Although N fertilizers have increased crop production, inefficient and excessive use has contributed to deterioration of the environment. The current rate of $N_r$ loss to the environment from global cropping systems is more than 10 times the rate that occurred at the end of 1800s (UNEP and WHRC, 2007). Significant amounts of N emissions in South Asia are produced from farming practices including manures and chemical fertilizer application as well as fossil fuel burning. In Pakistan, a massive increase in N surplus over time and resulting $N_r$ losses as $NH_3$, $N_2O$, and NO has been estimated (Khan et al., 2011; Raza et al., 2018). Total estimated N emissions from agriculture in Pakistan increased from 81 Gg N $yr^{-1}$ in 1961 to 1224 Gg N $yr^{-1}$ in 2013 (Raza et al., 2018).

$NH_3$, $N_2O$, and $NO_x$ are the major gaseous N losses in the $N_r$ cycle (Fowler et al., 2013). They can undergo chemical reactions in the atmosphere and impact global ecosystems thereby creating a cascading effect in the environment with serious consequences on air quality, terrestrial and aquatic ecosystems, biodiversity, food production, and human health (Liu et al., 2020; Fesenfeld et al., 2018). Fig. 6.2 depicts these complex processes and how $N_r$ can end up impacting our air, soil, and water quality through deposition and remobilization and chemical combination with other pollutants.

### 6.3.1 Impacts of $N_r$ emissions on terrestrial ecosystems and biodiversity loss

Atmospheric deposition of N ($NO_x$, $NH_3$) is a significant driver of biodiversity loss through acidification and eutrophication of both soils and aquatic systems. In aquatic systems, it contributes to $N_r$ enrichment of lakes, coastal waters, and open ocean

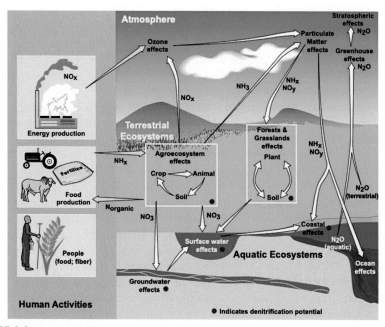

**FIGURE 6.2**

Cascading impact of reactive nitrogen in the environment.

*Reproduced from NEP and WHRC., 2007. Reactive Nitrogen in the Environment: Too Much or Too Little of a Good Thing; United Nations Environment Programme, Paris.*

leading to the proliferation of algal blooms and creation of an anoxic environment (Bobbink et al., 2010). The effect of N deposition can lead to both acidification and eutrophication of soils and surface waters, which in turn impacts species composition.

The adverse impact of N deposition is widespread and in order to address its reduction in a cost-effective manner, the concept of critical loads was introduced which takes into account the sensitivity of the ecosystem where deposition occurs. *Critical load* for N deposition is defined as the rate of N deposition (NO$_x$, NH$_y$) that can be sustained over a year, without causing significant harmful impact in sensitive habitats and ecosystems (Nilsson and Grennfelt, 1988; Rowe et al., 2019). Deposition above this limit is termed as *Critical Load Exceedance* and is typically mapped spatially to indicate the level of damage and the degree of intervention required.

NH$_3$ emissions are mostly deposited close to the source of emission through wet and dry deposition so the impact is usually very local. NH$_3$ together with NO$_x$ deposition is leading to a loss in plant diversity attributed to indirect soil mediated effects through N enrichment, eutrophication, and acidification (Bobbink et al., 2010). Results of various meta-analyses also indicate an overall decline in plant diversity in response to increased N availability and a shift toward more fast-growing N tolerant plants (Midolo et al., 2019; Ren et al., 2019). More N accumulation in foliage may increase the frequency of insect and pathogens attack and reduce frost hardiness. It also increases canopy size, shoot/root ratio, and loss of mychorrhizal infection, thus increasing the risk of drought stress (Erisman and Vries, 2000). Elevated N deposition is regarded as one of the important drivers of biodiversity loss. It can alter composition of community species or loss of plant species in terrestrial ecosystems. Nitrogen deposition in forests may enhance growth and productivity due to increased N availability during early stage. However, it may cause acidification and eutrophication at later stages thereby disturbing nutrient balance and increasing susceptibility to drought, pests, and diseases (Krupa, 2003).

$NH_3$ is alkaline; however, when it deposits on soil it causes acidification of the soil due to the nitrification processes where ammonium is oxidized to nitric acid (NAEI, 2008). Acidification of soil can in turn promote the leaching out of base cations such as calcium making the soil even more acidic and possibly leading to root damage and loss of species (DoE, 1994). Long-term high N deposition can influence nutrient imbalances in roots and leaves and cause soil acidification through releasing aluminum. Besides $NH_3$, deposition of $NO_x$ can also lead to acidification as it combines with moisture in the air to form nitric acid (Zhang et al., 2007). Although soil acidification is a natural and gradual process in soil development, high N use in agriculture has significantly increased soil acidification of croplands (Raza et al., 2021).

In Pakistan, N addition to croplands through atmospheric deposition during last six decades has been estimated to increase from 261 Gg N $yr^{-1}$ in 1961 to 589 in 2018 Gg N $yr^{-1}$ (FAOSTAT, 2021). As in several other regions, due to lack of well-designed monitoring systems, there are a few reports that have observed long-term changes in plant biodiversity in response to N deposition in Pakistan (Du et al., 2020).

### 6.3.2 Crop productivity and susceptibility

Direct exposure and uptake of elevated $NH_3$ concentration may cause damages to epicuticular wax layer in plants. It increases susceptibility to drought stress by accelerating stomatal opening and increasing transpiration rate. High $NH_3$ exposure enhances pest attacks and risks of fungal infection due to greater N concentration in plant tissues (Bobbink et al., 2010). Foliar uptake rate of $NH_3$ exceeding the detoxification rate induces direct adverse effects in plants. Acute toxicity of N gases and aerosols negatively impacts physiology and growth of the above ground plant parts of certain sensitive plants. Sensitivity to $NH_3$ exposure differs among plant species. For instance, epiphytic lichens are one of the most sensitive species to gaseous $NH_3$. The proposed average critical level of $NH_3$ for lichens and higher plants is 1 and 3 mg $NH_3$ $m^{-3}$, respectively (Cape et al., 2009).

$NO_x$ reacts with hydrocarbons in the presence of sunlight to form tropospheric ozone, $O_3$, which impacts crop health. It can also combine with moisture in the air to form nitric acid leading to acid precipitation (Zhang et al., 2007). Adverse impacts of tropospheric $O_3$ on terrestrial ecosystems include yield losses and reduced seed production. Severe effects in plants may occur when $O_3$ is absorbed through the stomata into leaf interior (Manning, 2005). After entering through leaf stomata, $O_3$ causes visible leaf injury, reduced photosynthesis, and ultimately growth and yield of crops (Feng et al., 2019). Li et al. (2017) reported in meta-analysis that high $O_3$ concentration (116 ppb) reduced total biomass of Chinese woody plants by 14% as compared with control groups (21 ppb). According to Yue et al. (2017), current $O_3$ level in China has caused a reduction in the annual net primary productivity by 10.1%–17.8%. Similarly, Feng et al. (2019) reported that current $O_3$ concentration in China has decreased rice yield by 8%, wheat yield by 6%, and forest tree biomass growth by 11%–13%. High $NO_x$ levels also lead to reduction in growth and yield of vegetation and agricultural crops. The global yield losses at 3%–16% in cereal crops were estimated in the year 2000 (Van Dingenen et al., 2009). In a meta-analysis, Wittig et al. (2009) found that mean $O_3$ concentration of 40 ppb caused reduction in total tree biomass by 7% increasing to 11% at 64 ppb.

In addition, through chemical reactions in the atmosphere, $NH_3$ can form $NH_4^+$ aerosols which in turn can disturb Earth's radiative balance acting as cloud condensation nuclei (Abbatt et al., 2006) and through scattering incoming radiations (Henze et al., 2012). The resulting haze pollution reduces the incoming solar radiation which can have a negative impact on regional climate, plant photosynthesis activity, and crop productivity (Zhou et al., 2018).

### 6.3.3 Impacts on human health

Increased $N_r$ contributes significantly to air pollution through the formation of tropospheric $O_3$ and secondary fine particulate matter ($PM_{2.5}$) (Liu et al., 2020). Both pollutants have an adverse impact on human health.

#### 6.3.3.1 PM2.5 exposure

High $NH_3$ concentration is one of the key precursors for aerosol formation in the atmosphere and may cause severe air pollution problems. It forms secondary inorganic aerosols by reacting with acidic components in the atmosphere, in particular $NO_x$ and $SO_4$ radicals (produced from fossil fuel combustion). These secondary particulates, $PM_{2.5}$, adversely impact human health and environment at a hemispherical scale. Its high level in the atmosphere has been linked to cardiovascular and respiratory ailments in humans, atmospheric haze production, and reduced visibility (Fu et al., 2015).

$PM_{2.5}$ negatively impacts human health by penetrating the respiratory systems and causing premature mortality (Lelieveld et al., 2015). In South Asia, $PM_{2.5}$ is linked to an 85% increase in premature mortality during 1990–2010 (Wang et al., 2017). Xu et al. (2018) considered agriculture as a major contributor to premature deaths in Asia because of $NH_3$ emissions from N fertilizer and livestock excreta.

#### 6.3.3.2 Tropospheric O3 exposure

Besides $PM_{2.5}$, $N_r$ is also a precursor in the formation of tropospheric $O_3$. $NO_x$ reacts with hydrocarbons to produce tropospheric $O_3$ which adversely impacts human health through inhalation and may lead to asthma (especially in children) and chronic respiratory disease. According to WHO (2008), $O_3$ negatively influences lung functions and increases mortality and respiratory morbidity rates. Lelieveld et al. (2015) estimated the global number of premature deaths in 2010 from exposure to tropospheric $O_3$ increased from 0.142 to 0.358 million between 2010 and 2050.

### 6.3.3.3 $NO_x$ exposure

Long-term exposure to elevated $NO_x$ concentration can cause chronic pulmonary and cardiovascular diseases. Impaired lung functions in adults has been documented due to high $NO_x$ exposure (Bowatte et al., 2017). High concentration of $NO_x$ can prolong and worsen common viral infections and may cause severe lungs damage (Spannhake et al., 2002). Many studies have reported associations between $NO_2$ exposure and hospital admissions with respiratory symptoms (WHO, 2013). Achakulwisut et al. (2019) estimated that $NO_2$ pollution is responsible for 4.0 million new pediatric asthma cases annually, sharing around 13% of the global asthma incidence.

### 6.3.3.4 UV radiation exposure

$N_2O$ emissions although fairly stable in the troposphere can be a significant contributor to destruction of stratospheric $O_3$ (Portmann et al., 2012), thereby increasing UV radiations and occurrence of skin disease cases. Besides being a powerful greenhouse gas with a global warming potential 265 times that of $CO_2$, $N_2O$ is the leading stratospheric ozone-depleting gas, and its concentration in the atmosphere is continuously increasing at the rate of 0.2%−0.3% per annum (IPCC, 2014).

## 6.4 Trends of N inputs and emissions

A wide disparity in rising synthetic N fertilizer consumption between developed and developing countries has been recorded during the past 5 decades (FAOSTAT, 2021). On global scale, developing countries are accounting for more than 50% of the total anthropogenic N. Global annual N inputs have been increased up to $\sim$174 Tg N $yr^{-1}$, with a mean N surplus of about 100 Tg. Among N inputs, synthetic fertilizers have major share (53%−59%), followed by livestock manure (16%−20%), biological N fixation (16%−18%), and atmospheric deposition (6%−10%). Asia, with 54% of the total global N input, has highest contribution among world's regions and is responsible for 67% of the global N surplus (Zhang et al., 2015). Streets et al. (2003) estimated about 27.5 Tg N $yr^{-1}$ as $NH_3$ emissions from all major anthropogenic sources in Asia, among which N fertilizer and animal waste contributed 45% and 38%, respectively. Ammonia emissions from N fertilizer and livestock manures in Southern Asia increased fourfold during the period from 1961 to 2014. Xu et al. (2018) estimated $21.3 \pm 3.9$ Tg N $yr^{-1}$ $NH_3$ emissions from agricultural systems in Southern Asia and reported a rapidly increasing rate of $\sim$0.3 Tg N $yr^{-1}$ during 1961−2014. Among emission sources, synthetic N fertilizer and manures contributed about 10.8 and 10.4 Tg N $yr^{-1}$, respectively.

After China and India, Pakistan has the highest N use and $NH_3$ emissions in the region. During 2000—14, N fertilizer application in Pakistan reached to 3.1 Tg N $yr^{-1}$ which contributed to $NH_3$ emissions of 716 Gg N $yr^{-1}$ (Xu et al., 2018), and $N_2O$ emissions of 58.9 Gg N $yr^{-1}$ (FAOSTAT, 2021). Likewise, amount of N from animal waste production during this period was 2.3 Tg N $yr^{-1}$ which resulted in $\sim$517 and 3.3 Gg N $yr^{-1}$ $NH_3$ and $N_2O$ emissions, respectively (Table 6.1). Van Damme et al. (2018) estimated global average $NH_3$ emissions over the period of 9 years. In total, 248 hotspots were identified with the largest average $NH_3$ column was found over the Indus Valley of Pakistan with $1.1 \times 10^{17}$ molecules $cm^{-1}$. Synthetic fertilizers and livestock are the major sources of $NH_3$ emissions in the region (Xu et al., 2018).

## 6.5 Nitrogen emissions from agriculture in Pakistan

### 6.5.1 N inputs

Pakistan has experienced a significant increase in N consumption in the agriculture sector. During the past six decades (1961—2018), increase in annual N input to croplands from different sources is indicated in Fig. 6.3. Almost 83% of the fertilizer consumed in Pakistan is domestically sourced from the various urea production facilities in the country (Raza et al., 2018).The amount of N in 1961 from synthetic fertilizers, manure applied to soils, biological N fixation, and atmospheric deposition was 42, 145, 51, and 261 Gg N $yr^{-1}$, respectively, reaching up to 3447, 786, 63, and 589 Gg N $yr^{-1}$, respectively, in 2018. Total N input from all the sources escalated from 499 to 4855 Gg N $yr^{-1}$ over this period (FAOSTAT, 2021). During this period, a substantial increase in the use of synthetic fertilizer was recorded whereas increase in manure applied to soils was considerably low. This does not seem to correspond to the total number of animal increases over the same period indicating either poor manure utilization or an incomplete understanding of manure usage in the different agricultural systems. Section 6.7 discusses this and other key information gaps.

### 6.5.2 N emissions

About 10-fold increase in total N input (including synthetic fertilizers, manure, biological N fixation, and atmospheric deposition) has been recorded in Pakistan between 1961 and 2018 (FAOSTAT, 2021). Raza et al. (2018) estimated a rapid increasing surplus N ranging from 171 to 3581 Gg N $yr^{-1}$ during the period 1961 to 2013. This has led to an increase in N emissions. According to FAOSTAT (2021), $NH_3$, $N_2O$, and NO emissions from synthetic fertilizer usage in Pakistan increased from 9.2, 1.0, and 0.6 Gg N $yr^{-1}$ in 1961 to 758.3, 86.2, and 51.7 Gg N $yr^{-1}$, respectively, in 2018 (Fig. 6.4).

The data regarding N emission resulting from manure application in Pakistan during last six decades (1961—2018) is depicted in Fig. 6.5. The $NH_3$, $N_2O$, and

**Table 6.1** Comparison among major countries of Asia for $NH_3$ and $N_2O$ emissions resulting from synthetic N fertilizer application and livestock excreta during 2000−14.

| Country | Synthetic N fertilizer | | | | Livestock excreta | | |
|---|---|---|---|---|---|---|---|
| | Application (Tg N yr$^{-1}$) | NH$_3$ emissions (Gg N yr$^{-1}$) | N$_2$O emissions (Gg N yr$^{-1}$) | Livestock excreta (Tg N yr$^{-1}$) | NH$_3$ emissions (Gg N yr$^{-1}$) | N$_2$O emissions (Gg N yr$^{-1}$) |
| China | 30.8 | 4148 | 575.6 | 16.2 | 3736 | 124.0 |
| India | 14.5 | 2840 | 293.1 | 6.2 | 1435 | 12.8 |
| Pakistan | 3.1 | 716 | 58.9 | 2.3 | 517 | 3.3 |
| Turkey | 1.4 | 143 | 28.3 | 1.8 | 406 | 0.6 |
| Iran | 1.2 | 105 | 19.4 | 1.8 | 406 | 2.2 |
| Indonesia | 2.9 | 365 | 52.3 | 1.0 | 228 | 12.5 |
| Vietnam | 1.4 | 253 | 24.9 | 0.8 | 188 | 8.3 |

Modified from Xu, R., Pan, S., Chen, J., Chen, G., Yang, J., Dangal, S., Shepard, J., Tian, H., 2018. Half-century ammonia emissions from agricultural systems in Southern Asia: magnitude, spatiotemporal patterns, and implications for human health. GeoHealth 2 (1), 40−53; FAOSTAT. 2021. Statistics Division (Rome: Food and Agriculture Organization of the United Nations) (www.fao.org/faostat/en/#data).

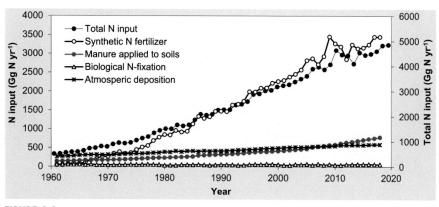

**FIGURE 6.3**

Historical changes in annual N input to croplands from different sources in Pakistan during last six decades (1961–2018).

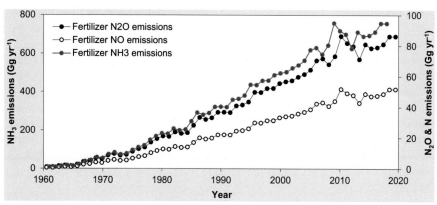

**FIGURE 6.4**

Historical changes in $NH_3$, $N_2O$, and NO emissions resulting from chemical fertilizers in Pakistan during last six decades (1961–2018).

NO emissions from manures in 1961 were 62.2, 7.8, and 1.2 Gg N $yr^{-1}$ which increased up to 338.3, 42.3, and 6.8 Gg N $yr^{-1}$, respectively, in 2018 (FAOSTAT, 2021).

## 6.6 Future projections
### 6.6.1 Global projections

The global population is projected to reach over nine billion by 2050, thus demanding about 70% and 100% increase in crop yields by developed and

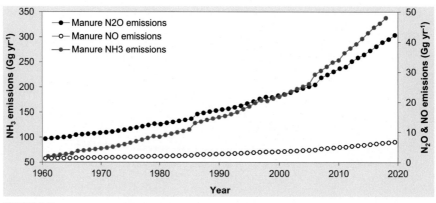

**FIGURE 6.5**

Historical changes in $NH_3$, $N_2O$, and NO emissions resulting from manure application in Pakistan during last six decades (1961–2018).

developing countries, respectively, to ensure food security (Rodriguez and Sanders, 2015). Therefore, more nutrient inputs, both organic and inorganic, will be consumed to support agricultural productivity. The global N cycle has experienced substantial changes due to continual application of N over the past five decades. Nitrogen applied in excess to crop demands is resulting significant N losses to the atmosphere as $NH_3$ and $N_2O$ emissions (Xu et al., 2020). Eickhout et al. (2006) projected a large increase in N loss (67–93 Tg N $yr^{-1}$) from developing countries between 1995 and 2030. Similarly, Vallack et al. (2001) predicted two to fourfold higher $NO_x$ emissions by the developing countries in 2025 than in 1995. The contribution in agricultural $NH_3$ emissions by developing countries increased from 63% to 76% during 1970 and 1995, and is predicted to rise to 80% by 2030 (Eickhout et al., 2006).

## 6.6.2 Regional projections

According to United Nations, Department of Economic and Social Affairs (2015), Asia is projected to be the second largest contributor to future population growth between 2005 and 2050. As a consequence, South and East Asia is expected to experience a rapid increase in both livestock products and crop foods as projected by FAO and other agricultural models (Valin et al., 2014). Therefore, manure production is expected to increase by 50% till 2050 as compared to 2000 (Bouwman et al., 2013). At a global level, total N demand for croplands under business as usual scenario is expected to double by 2100 (Stocker et al., 2013). Consequently, agricultural emissions are most likely to increase significantly (Xu et al., 2018). Taking the case of only paddy fields in Asia, $NH_3$ emissions are expected to increase from 13.8 Tg $yr^{-1}$ in 2000 to ~18.9 Tg $yr^{-1}$ by 2030 (Zheng et al., 2002).

### 6.6.3 Pakistan projections

According to FAOSTAT (2021), future projections regarding population, N input and N emissions in Pakistan is given in Table 6.2. Population of Pakistan in 2018 was 212 million which is projected to reach 338 million in 2050. Synthetic fertilizer and manure N inputs during 2018 were 3447 and 775 Gg, which are expected to increase to 6318 and 996 Gg by 2050, respectively. $NH_3$, $N_2O$, and NO emissions from fertilizers during 2018 were 758, 86, and 52 Gg which are predicted to increase up to 1390, 158, and 95 Gg, respectively, by 2050. Similarly, contribution of manures to $NH_3$, $N_2O$, and NO emissions during 2018 was estimated at 310, 86, and 6.2 Gg and projected to increase to 387, 48, and 7.7 Gg, respectively, by 2050.

## 6.7 Key information gaps

In order to develop a national strategy to reduce $N_r$ in the environment from agriculture, we need more data that is Pakistan specific and addresses the following key concerns.

### 6.7.1 Spatial distribution

The impact of $N_r$ can be local due to $NH_3$ deposition or transboundary such as $PM_{2.5}$ exposure, so understanding the spatial distribution and magnitude of the sources is critical. Although data on animals, fertilizer use, and related N emissions are available at the national level and indicate a significant increase in fertilizer use and estimated N emissions between 1961 and 2018, it is difficult to ascertain how this increase is distributed across the country. Finer spatial resolution data are required to better understand the distribution of fertilizer usage, animal numbers, cropping, and grazing activities. Moreover, there are little data on rangelands and grasslands which are in the rain-fed parts of the country and where livestock rearing plays a significant role in the economy while cropping systems may be different to those in the irrigated regions. The FAOSTAT data used in this report emphasize this gap in data. At a national level, there has been an 82-fold increase in N inputs from artificial fertilizer usage between 1961 and 2018 (42−3447 Gg N $yr^{-1}$), whereas the corresponding increase in N inputs from manure applied to soils is only 5.4 (145−786 Gg N $yr^{-1}$) (Figs. 6.1 and 6.3). This small increase in manure usage does not correspond to the significant increase in animal numbers, 268% for grazing livestock and 2597% for poultry, over the same period (Raza et al., 2018; FAOSTAT, 2021). So better understanding of the spatial distribution of animals and manure application is key to developing an effective strategy to reduce $N_r$ in the environment.

### 6.7.2 System-based analysis

More data are also required to understand and document the many cropping and livestock practices in the country using a systems-based approach. This will allow a

**Table 6.2** Future projections for population, N input, and N emissions in Pakistan.

| Year | Population (millions) | N input (Gg N yr$^{-1}$) | | NH$_3$ emissions (Gg N yr$^{-1}$) | | N$_2$O emissions (Gg N yr$^{-1}$) | | NO emissions (Gg N yr$^{-1}$) | |
|------|------|-----------|--------|-----------|--------|-----------|--------|-----------|--------|
| | | Fertilizer | Manure | Fertilizer | Manure | Fertilizer | Manure | Fertilizer | Manure |
| 2018 | 212 | 3447 | 775 | 758 | 310 | 86 | 39 | 52 | 6.2 |
| 2030 | 263 | 5021 | 846 | 1105 | 338 | 126 | 42 | 75 | 6.8 |
| 2050 | 338 | 6318 | 966 | 1390 | 387 | 158 | 48 | 95 | 7.7 |

*Modified from FAOSTAT. 2021. Statistics Division (Rome: Food and Agriculture Organization of the United Nations) (www.fao.org/faostat/en/#data).*

better understanding of the N pathways, N utilization, and the potential for reducing N losses from each system. Besides the varied cropping and livestock systems, the size of the average farm is fairly small in Pakistan. In the case of animals, though there has been an increase in feedlots and poultry rearing units, these are not the norm. Many households keep a few animals to either supplement nutritional needs or use as a cashable asset. There is a need for more information on how N from manure is managed by these households. It also means that the fertilizer application, manure management solutions, and the measures to reduce N emissions being advocated in Europe and other more industrialized parts of the world may not be directly transferable or affordable for an average farmer in Pakistan. There are also gaps in knowledge of more traditional tried and tested practices developed over centuries that are low input and utilize manure/natural fertilizers. Including these to inform future strategies has the potential to provide more contextual and culturally acceptable solutions.

### 6.7.3 Rangelands management

Rangelands form over 60% of the country with the main form of agriculture centered around livestock systems. Subsistence cropping practices, with low N inputs derived from legumes or manure, are used. Encroachment of more intensive forms of cropping in these regions is depleting water resources, increasing pollution, and making the communities more vulnerable to climate change due to degrading resources and more exposure to market volatility. To make matters worse, current grazing practices of under-impacting land and overgrazing plants by animals are not sustainable, leading to further land degradation. Conventional agricultural approaches, which put a high demand on natural resources, are not suitable in these terrains without modification. There is a need to determine the impact of these conventional practices on the environment and communities. In addition, there are gaps in our understanding of the empirical and local knowledge held by local communities, their livestock management and grazing practices, cropping systems, and the N pathways for these systems. These are important to understand before bringing in management strategies to improve grazing practices and cropping systems in these marginal landscapes.

### 6.7.4 Nitrogen utilization

Crop needs, timing, and proper application need to be better understood in Pakistan's context. There is a general perception that soils in Pakistan are N deficient and so often the advice given by agricultural extension teams and agri-consultants favors application of N fertilizers. Little is documented on how effective and appropriate this advice is especially in relation to crop needs, timing, and environmental impact. The link between low N utilization efficiency and low soil organic carbon needs to be better understood in addition to the processes that are depleting soil organic carbon in Pakistan. Moreover, good practices such as introduction of legumes, crop diversity, cover crops, livestock systems that are grass-fed and not requiring high input fodder crop, together with nature friendly soil management practices, need to be investigated within Pakistan's landscape. The impact of these approaches on enhancing soil biology to make key minerals locked in soil bioavailable to the plant needs to be better understood. This is a popular approach being adopted in many parts of the world to reduce fertilizer inputs, protect soil, enhance productivity and increase farmer profits.

### 6.7.5 Emission factors

The emission factors used for estimating N emissions, though a good starting point, have not been developed specifically for Pakistan, so do not capture the ambient conditions which play an important part in determining their value. The aforementioned knowledge gap in understanding the local agricultural systems, the N pathways need to be plugged which together with application of local environmental conditions can help move toward development of locally applicable emission factors. More research and participatory trials are needed to refine this information for the Pakistan context.

### 6.7.6 Data on impact

Another gap is in understanding the impact of $N_r$ on health, cropping systems, and the environment in Pakistan. Poor data availability, coupled with the lack of studies that specifically look at links between $N_r$ and impacts, is a likely reason. Moreover, there is limited information on sensitive ecosystems and their vulnerability to N deposition. There is a need to have a better understanding of the distribution and magnitude of nonagricultural sources of $N_r$ emissions such as fossil fuel burning, so as to model the interactions with agricultural emissions and determine likely impact.

### 6.7.7 Food waste and supply chain

Data on postharvest losses across the supply chain from production to consumption are limited. Understanding where the losses are will allow developing strategies to reduce these losses and in turn reduce the loss of $N_r$ at each stage from field to fork.

### 6.7.8 Policy instruments, financial costs, and benefits

One of the key policy instruments that have led to the increase in use of synthetic fertilizers over natural ones has been the introduction of agricultural subsidies. It not only distorts the market value but when combined with the implicit exclusion of externalities (negative impact on environment and health) in the price, makes the use of synthetic fertilizers more profitable, especially in the short term. There is a gap in understanding the impact of subsidies, the hidden cost of excessive $N_r$ on ecosystem services and human health, hence the true value of protecting natural capital and ecosystem services. Understanding the wider environmental and health benefits and monetizing these benefits is a key starting point to assessing strategies that reduce $N_r$ in the environment.

## 6.8 **Summary and next steps required**

Nitrogen has been one of the most limiting nutrients in agriculture, thus requiring enhanced N cycling to uphold crop productivity. N emissions due to high N use in agriculture adversely impact terrestrial ecosystems and human health; however, crop production without N is generally not considered to be an option to meet future population needs. Applying more N, in excess crop needs, results in excess $N_r$ in the environment. This then undergoes a series of complex transformations and reactions with other pollutants adversely impacting aquatic systems, biodiversity, and human health and reducing crop productivity.

For Pakistan, although a significant increase in the use of synthetic fertilizers as well as livestock numbers is documented at the national level, there is a need to refine this information in key areas: N usage, spatial distribution of livestock and cropping systems, N pathways within these systems, Pakistan-specific emission factors, the impact on the environment and health, losses across the supply chain, and finally financial cost and benefits of reducing $N_r$ in the different agricultural systems. These have been discussed in more detail in Section 6.7.

The way forward is to close these information gaps through collection of appropriate data and establishing the baseline situation for N usage, its distribution and practice, and the impacts. Therefore, this issue should be addressed collaboratively with a wide group of stakeholders including policy makers, farming communities, agronomists, research bodies, agri-extension support, suppliers, and key stakeholders within the supply chain. Nature-based and cost-effective strategies are urgently required to improve N use efficiency, reduce $N_r$ and address environmental problems without yield penalties.

Nitrogen overuse and emissions are among the most pressing environmental issues that the world is facing. But in spite of the enormity of our influence on the N cycle and resulting implications for the environment and public health, there is still little attention being paid to the issue. More than half of the N applied to agricultural soils is lost to the environment. Effective, integrated farming strategies should be implemented that reduce demand for N and limit $N_r$ losses from agricultural lands, thus mitigating associated threats to human health and environment while improving farm profitability. These strategies also need to be optimized across the supply chain from production to consumption.

### 6.8.1 **Next steps**

To address the $N_r$ issue comprehensively, there is a need to develop policy and agricultural strategies that not only take account of the wider impact of excess $N_r$ on the environment (pollution, biodiversity loss, health risk, climate change) but also address the broader issues in the supply chain (e.g., food availability, food loss, and lack of market accessibility). Understanding the different agricultural systems for both cropping and livestock management in the country, the local terrain and climate, the complexity within which farming communities works is therefore imperative to develop realistic solutions that have win-wins for reducing emissions, pollution, and greenhouse gases while improving productivity and farmer profitability.

In addition, it is important to identify and then monitor the N usage and losses in each of these systems, map them spatially to determine possible impact on the environment and on public health. Few baseline assessments of soil health, water quality, and air pollution mean there is little understanding of the impact of $N_r$ on the environment, health, and agricultural productivity. More research is required to increase this knowledge base and to find appropriate solutions. There is also a need to develop N usage, application, and monitoring solutions that are appropriate for the local context in Pakistan as farms are generally much smaller and ownership and management practices vary from those found in Europe or United States where many of the $N_r$ studies have been carried out.

There is a need to understand the economic and environmental impact of low input–low pollution cropping and livestock systems. These could include the use of leguminous crops, cover cropping, inhibitors for cropping systems, or planned extensive, rotational grazing in the livestock sector. There are existing traditional systems that have relied upon low N inputs and these need to be revisited not only to learn from them but also to make them more effective where necessary.

In addition, agri-extension staff and agronomists need capacity building training on best practices for fertilizer and manure use and how to implement it within a given context. The training would need to include understanding the environmental and health impact of N losses in the system. Extending this knowledge to farmers would help them not only utilize N more effectively but also recognize the benefits to their health, environment, and profits. Finally, creating appropriate financial instruments and technical support systems that help farmers transition to low N-input systems as well as reduce the financial risk of climatic and market variability need to be part of a future, more holistic N strategy.

## References

Abbatt, J., Benz, S., Cziczo, D., Kanji, Z., Lohmann, U., Möhler, O., 2006. Solid ammonium sulfate aerosols as ice nuclei: a pathway for cirrus cloud formation. Science 313 (5794), 1770–1773.

Abrar, M.M., Xu, H., Aziz, T., Sun, N., Mustafa, A., Aslam, M.W., Shah, S.A.A., Mehmood, K., Zhou, B., Ma, X., 2020. Carbon, nitrogen, and phosphorus stoichiometry mediate sensitivity of carbon stabilization mechanisms along with surface layers of a Mollisol after long-term fertilization in Northeast China. J. Soils Sediments 21 (2), 705–723.

Achakulwisut, P., Brauer, M., Hystad, P., Anenberg, S.C., 2019. Global, national, and urban burdens of paediatric asthma incidence attributable to ambient $NO_2$ pollution: estimates from global datasets. Lancet Planet. Health 3 (4), e166–e178.

Ashraf, M.N., Hu, C., Wu, L., Duan, Y., Zhang, W., Aziz, T., Cai, A., Abrar, M.M., Xu, M., 2020. Soil and microbial biomass stoichiometry regulate soil organic carbon and nitrogen mineralization in rice-wheat rotation subjected to long-term fertilization. J. Soils Sediments 20 (8), 3103–3113.

Barnes, I., Rudzinski, K.J., 2006. Environmental Simulation Chambers: Application to Atmospheric Chemical Processes, vol. 62. Springer Science & Business Media.

Bobbink, R., Hicks, K., Galloway, J., Spranger, T., Alkemade, R., Ashmore, M., Bustamante, M., Cinderby, S., Davidson, E., Dentener, F., 2010. Global assessment of nitrogen deposition effects on terrestrial plant diversity: a synthesis. Ecol. Appl. 20 (1), 30–59.

Bouwman, L., Goldewijk, K.K., Van Der Hoek, K.W., Beusen, A.H., Van Vuuren, D.P., Willems, J., Rufino, M.C., Stehfest, E., 2013. Exploring global changes in nitrogen and phosphorus cycles in agriculture induced by livestock production over the 1900–2050 period. Proc. Natl. Acad. Sci. U.S.A. 110 (52), 20882–20887.

Bowatte, G., Erbas, B., Lodge, C.J., Knibbs, L.D., Gurrin, L.C., Marks, G.B., Thomas, P.S., Johns, D.P., Giles, G.G., Hui, J., 2017. Traffic-related air pollution exposure over a 5-year period is associated with increased risk of asthma and poor lung function in middle age. Eur. Respir. J. 50 (4).

Brady, N.C., Weil, R.R., 2012. The nature and properties of soils. In: Nitrogen and Sulfur Economy of Soil, vol. 14. Prentice Hall, Upper Saddle River, NJ, pp. 542–593.

Cameron, K.C., Di, H.J., Moir, J.L., 2013. Nitrogen losses from the soil/plant system: a review. Ann. Appl. Biol. 162 (2), 145–173.

Cape, J., Van der Eerden, L., Sheppard, L., Leith, I., Sutton, M., 2009. Evidence for changing the critical level for ammonia. Environ. Pollut. 157 (3), 1033–1037.

Chen, A., Lei, B., Hu, W., Lu, Y., Mao, Y., Duan, Z., Shi, Z., 2015. Characteristics of ammonia volatilization on rice grown under different nitrogen application rates and its quantitative predictions in Erhai Lake Watershed, China. Nutrient Cycl. Agroecosyst. 101 (1), 139–152.

Chen, D., Pan, Q., Bai, Y., Hu, S., Huang, J., Wang, Q., Naeem, S., Elser, J.J., Wu, J., Han, X., 2016. Effects of plant functional group loss on soil biota and net ecosystem exchange: a plant removal experiment in the Mongolian grassland. J. Ecol. 104 (3), 734–743.

Crowther, T.W., Thomas, S.M., Maynard, D.S., Baldrian, P., Covey, K., Frey, S.D., van Diepen, L.T., Bradford, M.A., 2015. Biotic interactions mediate soil microbial feedbacks to climate change. Proc. Natl. Acad. Sci. U.S.A. 112 (22), 7033–7038.

Dangal, S.R., Tian, H., Zhang, B., Pan, S., Lu, C., Yang, J., 2017. Methane emission from global livestock sector during 1890–2014: magnitude, trends and spatiotemporal patterns. Global Change Biol. 23 (10), 4147–4161.

DoE, 1994. Impacts of nitrogen deposition in terrestrial systems. In: The United Kingdom Review Group on Impacts of Atmospheric Nitrogen, Prepared for the Department of the Environment.

Du, E., Lu, X., Tian, D., Mao, Q., Jing, X., Wang, C., Xia, N., 2020. Impacts of nitrogen deposition on forest ecosystems in China. In: Atmospheric Reactive Nitrogen in China. Springer, pp. 185–213.

Eickhout, B., Bouwman, A.V., Van Zeijts, H., 2006. The role of nitrogen in world food production and environmental sustainability. Agric. Ecosyst. Environ. 116 (1–2), 4–14.

Erisman, J.W., Vries, W.d., 2000. Nitrogen deposition and effects on European forests. Environ. Rev. 8 (2), 65–93.

FAOSTAT, 2021. Statistics Division (Rome: Food and Agriculture Organization of the United Nations). www.fao.org/faostat/en/#data.

Feng, Z., De Marco, A., Anav, A., Gualtieri, M., Sicard, P., Tian, H., Fornasier, F., Tao, F., Guo, A., Paoletti, E., 2019. Economic losses due to ozone impacts on human health, forest productivity and crop yield across China. Environ. Int. 131, 104966.

Fesenfeld, L.P., Schmidt, T.S., Schrode, A., 2018. Climate policy for short-and long-lived pollutants. Nat. Clim. Change 8 (11), 933–936.

Fowler, D., Steadman, C.E., Stevenson, D., Coyle, M., Rees, R.M., Skiba, U., Sutton, M., Cape, J.N., Dore, A., Vieno, M., 2015. Effects of global change during the 21st century on the nitrogen cycle. Atmos. Chem. Phys. 15 (24), 13849–13893.

Fowler, D., Coyle, M., Skiba, U., Sutton, M.A., Cape, J.N., Reis, S., Sheppard, L.J., Jenkins, A., Grizzetti, B., Galloway, J.N., 2013. The global nitrogen cycle in the twenty-first century. Phil. Trans. Biol. Sci. 368 (1621), 20130164.

Fu, X., Wang, S., Ran, L., Pleim, J., Cooter, E., Bash, J., Benson, V., Hao, J., 2015. Estimating $NH_3$ emissions from agricultural fertilizer application in China using the bi-directional CMAQ model coupled to an agro-ecosystem model. Atmos. Chem. Phys. 15 (12), 6637–6649.

Ghaly, A., Ramakrishnan, V., 2015. Nitrogen sources and cycling in the ecosystem and its role in air, water and soil pollution: a critical review. J. Pollut. Effects Control 1–26.

Henze, D.K., Shindell, D.T., Akhtar, F., Spurr, R.J., Pinder, R.W., Loughlin, D., Kopacz, M., Singh, K., Shim, C., 2012. Spatially refined aerosol direct radiative forcing efficiencies. Environ. Sci. Technol. 46 (17), 9511–9518.

Huang, R.-J., Zhang, Y., Bozzetti, C., Ho, K.-F., Cao, J.-J., Han, Y., Daellenbach, K.R., Slowik, J.G., Platt, S.M., Canonaco, F., 2014. High secondary aerosol contribution to particulate pollution during haze events in China. Nature 514 (7521), 218–222.

IPCC, 2014. Climatechange2014: Impacts, Adaptation and Vulnerability. IPCC Working Group II Contribution to AR5. IPCC, Geneva, Switzerland. www.ipcc-wg2.gov/AR5/.

Irfan, M., Abbas, M., Shah, J.A., Memon, M.Y., 2018. Grain yield, nutrient accumulation and fertilizer efficiency in bread wheat under variable nitrogen and phosphorus regimes. J. Basic Appl. Sci. 14, 80–86.

Khan, A.N., Ghauri, B., Jilani, R., Rahman, S., 2011. Climate change: emissions and sinks of greenhouse gases in Pakistan. In: Proceedings of the Symposium on Changing Environmental Pattern and its Impact with Special Focus on Pakistan.

Krupa, S., 2003. Effects of atmospheric ammonia ($NH_3$) on terrestrial vegetation: a review. Environ. Pollut. 124 (2), 179–221.

Leininger, S., Urich, T., Schloter, M., Schwark, L., Qi, J., Nicol, G.W., Prosser, J.I., Schuster, S., Schleper, C., 2006. Archaea predominate among ammonia-oxidizing prokaryotes in soils. Nature 442 (7104), 806–809.

Lelieveld, J., Evans, J.S., Fnais, M., Giannadaki, D., Pozzer, A., 2015. The contribution of outdoor air pollution sources to premature mortality on a global scale. Nature 525 (7569), 367–371.

Li, P., Feng, Z., Catalayud, V., Yuan, X., Xu, Y., Paoletti, E., 2017. A meta-analysis on growth, physiological, and biochemical responses of woody species to ground-level ozone highlights the role of plant functional types. Plant Cell Environ. 40 (10), 2369–2380.

Liu, J., You, L., Amini, M., Obersteiner, M., Herrero, M., Zehnder, A.J., Yang, H., 2010. A high-resolution assessment on global nitrogen flows in cropland. Proc. Natl. Acad. Sci. U.S.A. 107 (17), 8035–8040.

Liu, X., Xu, W., Du, E., Tang, A., Zhang, Y., Zhang, Y., Wen, Z., Hao, T., Pan, Y., Zhang, L., et al., 2020. Environmental impacts of nitrogen emissions in China and the role of policies in emission reduction. Philos. Trans. R. Soc. A 378 (2183), 20190324.

Lu, X., Zhang, L., Chen, Y., Zhou, M., Zheng, B., Li, K., Liu, Y., Lin, J., Fu, T.-M., Zhang, Q., 2019. Exploring 2016–2017 surface ozone pollution over China: source contributions and meteorological influences. Atmos. Chem. Phys. 19 (12), 8339–8361.

Manning, W.J., 2005. Establishing a cause and effect relationship for ambient ozone exposure and tree growth in the forest: progress and an experimental approach. Environ. Pollut. 137 (3), 443–454.

Maqsood, M.A., Awan, U.K., Aziz, T., Arshad, H., Ashraf, N., Ali, M., 2016. Nitrogen management in calcareous soils: problems and solutions. Pakistan J. Agric. Sci. 53 (1), 79–95.

Marschner, H., 2012. Mineral Nutrition of Higher Plants, fifth ed. Academic Press, London. https://doi.org/10.1016/C2009-0-63043-9.

Martínez-Dalmau, J., Berbel, J., Ordóñez-Fernández, R., 2021. Nitrogen fertilization. A review of the risks associated with the inefficiency of its use and policy responses. Sustainability 13 (10), 5625.

Midolo, G., Alkemade, R., Schipper, A.M., Benítez-López, A., Perring, M.P., De Vries, W., 2019. Impacts of nitrogen addition on plant species richness and abundance: a global meta-analysis. Global Ecol. Biogeograp. 28 (3), 398–413.

NAEI, 2008. UNECE Emission Estimates to 2005 for Ammonia. National Air Emission Inventory, United Kingdom.

Nilsson, J., Grennfelt, P., 1988. Critical Loads for Sulphur and Nitrogen. Report from Skokloster Workshop. Skokloster. https://doi.org/10.1007/978-94-009-4003-1_11.

Portmann, R., Daniel, J., Ravishankara, A., 2012. Stratospheric ozone depletion due to nitrous oxide: influences of other gases. Phil. Trans. Biol. Sci. 367 (1593), 1256–1264.

Quévreux, P., Barot, S., Thébault, É., 2018. Impact of nutrient cycling on food web stability. bioRxiv 276592.

Rashid, A., Memon, K.S., 2010. Soil science. In: Soil and Fertilizer Nitrogen. National Book Foundation, Islamabad, Pakistan, pp. 261–289.

Raza, S., Zhou, J., Aziz, T., Afzal, M.R., Ahmed, M., Javaid, S., Chen, Z., 2018. Piling up reactive nitrogen and declining nitrogen use efficiency in Pakistan: a challenge not challenged (1961–2013). Environ. Res. Lett. 13 (3), 034012.

Raza, S., Zamanian, K., Ullah, S., Kuzyakov, Y., Virto, I., Zhou, J., 2021. Inorganic carbon losses by soil acidification jeopardize global efforts on carbon sequestration and climate change mitigation. J. Clean. Prod. 128036.

Ren, C., Liu, S., Van Grinsven, H., Reis, S., Jin, S., Liu, H., Gu, B., 2019. The impact of farm size on agricultural sustainability. J. Clean. Prod. 220, 357–367.

Riddick, S., Ward, D., Hess, P., Mahowald, N., Massad, R., Holland, E., 2016. Estimate of changes in agricultural terrestrial nitrogen pathways and ammonia emissions from 1850 to present in the Community Earth System Model. Biogeosciences 13 (11), 3397–3426.

Rodriguez, A., Sanders, I.R., 2015. The role of community and population ecology in applying mycorrhizal fungi for improved food security. ISME J. 9 (5), 1053–1061.

Rotz, C., 2004. Management to reduce nitrogen losses in animal production. J. Anim. Sci. 82 (Suppl. 1_13), E119–E137.

Rotz, C.A., Leytem, A.B., 2015. Reactive nitrogen emissions from agricultural operations. EM Mag. Air Waste Manag. Assoc. 12–17.

Rotz, C.A., Montes, F., Hafner, S.D., Heber, A.J., Grant, R.H., 2014. Ammonia emission model for whole farm evaluation of dairy production systems. J. Environ. Qual. 43 (4), 1143–1158.

Rousk, J., Bååth, E., Brookes, P.C., Lauber, C.L., Lozupone, C., Caporaso, J.G., Knight, R., Fierer, N., 2010. Soil bacterial and fungal communities across a pH gradient in an arable soil. ISME J. 4 (10), 1340–1351.

Rowe, E., Sawicka, K., Mitchell, Z., Smith, R., Dore, T., Banin, L.F., Levy, P., 2019. Trends Report 2019: Trends in Critical Load and Critical Level Exceedances in the UK. Report to Defra under Contract AQ0843, CEH Project NEC05708. https://uk-air.defra.gov.uk/library/.

Sarabia, L., Solorio, F.J., Ramírez, L., Ayala, A., Aguilar, C., Ku, J., Almeida, C., Cassador, R., Alves, B.J., Boddey, R.M., 2020. Improving the nitrogen cycling in livestock systems through silvopastoral systems. In: Nutrient Dynamics for Sustainable Crop Production. Springer, pp. 189–213.

Shahzad, A.N., Qureshi, M.K., Wakeel, A., Misselbrook, T., 2019. Crop production in Pakistan and low nitrogen use efficiencies. Nat. Sustain. 2 (12), 1106–1114.

Spannhake, E.W., Reddy, S.P., Jacoby, D.B., Yu, X.-Y., Saatian, B., Tian, J., 2002. Synergism between rhinovirus infection and oxidant pollutant exposure enhances airway epithelial cell cytokine production. Environ. Health Perspect. 110 (7), 665–670.

Steinfeld, H., Gerber, P., Wassenaar, T., Castel, V., Rosales, M., Rosales, M., de Haan, C., 2006. Livestock's long shadow: environmental issues and options. In: Livestock's Role in Climate Change and Air Pollution. LEAD/FAO, Rome, Italy, pp. 79–111.

Stocker, B.D., Roth, R., Joos, F., Spahni, R., Steinacher, M., Zaehle, S., Bouwman, L., Prentice, I.C., 2013. Multiple greenhouse-gas feedbacks from the land biosphere under future climate change scenarios. Nat. Clim. Change 3 (7), 666–672.

Streets, D.G., Bond, T., Carmichael, G., Fernandes, S., Fu, Q., He, D., Klimont, Z., Nelson, S., Tsai, N., Wang, M.Q., 2003. An inventory of gaseous and primary aerosol emissions in Asia in the year 2000. J. Geophys. Res. Atmosp. 108 (D21).

Sutton, M.A., Howard, C.M., Kanter, D.R., Lassaletta, L., Móring, A., Raghuram, N., Read, N., 2021. The nitrogen decade: mobilizing global action on nitrogen to 2030 and beyond. One Earth 4 (1), 10–14.

Sutton, M.A., Reis, S., Riddick, S.N., Dragosits, U., Nemitz, E., Theobald, M.R., Tang, Y.S., Braban, C.F., Vieno, M., Dore, A.J., 2013. Towards a climate-dependent paradigm of ammonia emission and deposition. Phil. Trans. Biol. Sci. 368 (1621), 20130166.

Tian, H., Banger, K., Bo, T., Dadhwal, V.K., 2014. History of land use in India during 1880–2010: large-scale land transformations reconstructed from satellite data and historical archives. Global Planet. Change 121, 78–88.

UNEP and WHRC, 2007. Reactive Nitrogen in the Environment: Too Much or Too Little of a Good Thing. United Nations Environment Programme, Paris.

United Nations, Department of Economic and Social Affairs, Population Division, 2015. World Population Prospects: The 2015 Revision, Key Findings and Advance Tables (Working Paper No. ESA/P/WP.241). Retrieved from. https://esa.un.org/unpd/wpp/publications/files/key_findings_wpp_2015.pdf.

Valin, H., Sands, R.D., Van der Mensbrugghe, D., Nelson, G.C., Ahammad, H., Blanc, E., Bodirsky, B., Fujimori, S., Hasegawa, T., Havlik, P., 2014. The future of food demand: understanding differences in global economic models. Agric. Econ. 45 (1), 51–67.

Vallack, H.W., Cinderby, S., Kuylenstierna, J.C., Heaps, C., 2001. Emission inventories for $SO_2$ and $NO_x$ in developing country regions in 1995 with projected emissions for 2025 according to two scenarios. Water Air Soil Pollut. 130 (1), 217–222.

Van Damme, M., Clarisse, L., Whitburn, S., Hadji-Lazaro, J., Hurtmans, D., Clerbaux, C., Coheur, P.-F., 2018. Industrial and agricultural ammonia point sources exposed. Nature 564 (7734), 99–103.

Van Dingenen, R., Dentener, F.J., Raes, F., Krol, M.C., Emberson, L., Cofala, J., 2009. The global impact of ozone on agricultural crop yields under current and future air quality legislation. Atmos. Environ. 43 (3), 604–618.

Vanlauwe, B., Diels, J., Aihou, K., Iwuafor, E., Lyasse, O., Sanginga, N., Merckx, R., 2002. Direct interactions between N fertilizer and organic matter: evidence from trials with 15N-labelled fertilizer. In: Integrated Plant Nutrient Management in Sub-saharan Africa: From Concept to Practice. CAB International, Wallingford, UK, pp. 173–184.

von Grebmer, K., et al., 2018. Global Hunger Index (Welthungerhilfe and Concern Worldwide, 2018). https://go.nature.com/2Q7JO1e.

Wang, J., Xing, J., Mathur, R., Pleim, J.E., Wang, S., Hogrefe, C., Gan, C.-M., Wong, D.C., Hao, J., 2017. Historical trends in PM2. 5-related premature mortality during 1990–2010 across the northern hemisphere. Environ. Health Perspect. 125 (3), 400–408.

WHO, 2008. Health risks of ozone form long-range transboundary air pollution. In: Report of World Health Organisation; 2008. Regional Office for Europe, Copenhagen, Denmark.

WHO, 2013. Health Risks of Air Pollution in Europe - HRAPIE Project: New Emerging Risks to Health from Air Pollution - Results from the Survey of Experts. World Health Organization, Regional office for Europe, Copenhagen.

Wittig, V.E., Ainsworth, E.A., Naidu, S.L., Karnosky, D.F., Long, S.P., 2009. Quantifying the impact of current and future tropospheric ozone on tree biomass, growth, physiology and biochemistry: a quantitative meta-analysis. Global Change Biol. 15 (2), 396–424.

Xu, P., Chen, A., Houlton, B.Z., Zeng, Z., Wei, S., Zhao, C., Lu, H., Liao, Y., Zheng, Z., Luan, S., 2020. Spatial variation of reactive nitrogen emissions from China's croplands codetermined by regional urbanization and its feedback to global climate change. Geophys. Res. Lett. 47 (12) e2019GL086551.

Xu, R., Pan, S., Chen, J., Chen, G., Yang, J., Dangal, S., Shepard, J., Tian, H., 2018. Half-century ammonia emissions from agricultural systems in Southern Asia: magnitude, spatio-temporal patterns, and implications for human health. GeoHealth 2 (1), 40–53.

Yue, X., Unger, N., Harper, K., Xia, X., Liao, H., Zhu, T., Xiao, J., Feng, Z., Li, J., 2017. Ozone and haze pollution weakens net primary productivity in China. Atmos. Chem. Phys. 17 (9), 6073–6089.

Zhang, Q., Streets, D.G., He, K., Wang, Y., Richter, A., Burrows, J.P., Uno, I., Jang, C.J., Chen, D., Yao, Z., 2007. NOx emission trends for China, 1995–2004: the view from the ground and the view from space. J. Geophys. Res. Atmosp. 112 (D22).

Zhang, X., Davidson, E.A., Mauzerall, D.L., Searchinger, T.D., Dumas, P., Shen, Y., 2015. Managing nitrogen for sustainable development. Nature 528 (7580), 51–59.

Zheng, X., Fu, C., Xu, X., Yan, X., Huang, Y., Han, S., Hu, F., Chen, G., 2002. The Asian nitrogen cycle case study. Ambio 79–87.

Zhou, L., Chen, X., Tian, X., 2018. The impact of fine particulate matter (PM2. 5) on China's agricultural production from 2001 to 2010. J. Clean. Prod. 178, 133–141.

# Nitrogen use efficiency in crop production: issues and challenges in South Asia

**Bijay-Singh[1], Hafiz Muhammad Bilal[2], Tariq Aziz[3]**

[1]*Punjab Agricultural University, Ludhiana, Punjab, India;* [2]*Water Management Research Farm, Renala Khurd, Okara, Pakistan;* [3]*University of Agriculture Faisalabad, Sub-Campus at Depalpur, Okara, Punjab, Pakistan*

## 7.1 Introduction

The eight countries in South Asia—Afghanistan, Bangladesh, Bhutan, India, Maldives, Nepal, Pakistan, and Sri Lanka—are home to about 1.8 billion people or 23.1% of the global population (https://www.worldometers.info/population/, accessed 28 May 2021). Agriculture is the mainstay of economy and social development in the region. The Green Revolution in South Asia fueled by high-yielding varieties of rice, wheat, and maize and increased use of fertilizer nitrogen (N) helped produce enough food grains to feed the population with the highest growth rate. Fertilizers were introduced in South Asia along with the Green Revolution package during the 1960s. Initially, the farmers were hesitant to use fertilizers. But soon after realizing the high yield gains resulting from the application of fertilizers, particularly N, fertilizer use increased rapidly. In 2018, the farmers in South Asia used 21.63% of the total fertilizer N consumed in the world (http://www.fao.org/faostat/en/#data/RFN, accessed 27 May 2021). Of the 22.433 million tonnes (Mt) of fertilizer N consumed in South Asia, 5.89%, 78.63%, and 14.56% was used in Bangladesh, India, and Pakistan, respectively. However, as shown in Fig. 7.1, time trends in fertilizer N use per unit of arable land in the three countries presented a different picture. After the mid-1980s, farmers in Bangladesh on average started applying higher fertilizer N rates to crops than those in India and Pakistan.

The proportion of applied N utilized by the crop plants is termed nitrogen use efficiency (NUE). Ideally, high NUE is observed when fertilizer N is applied in levels, which supplement soil N to produce optimum yields. On the contrary, when fertilizer N is not adequately managed or applied in heavy doses, resulting NUE is low and a portion of N may be lost from the soil—plant system and may affect the environment leading to nitrate pollution of surface and groundwater bodies (Bijay-Singh and Craswell, 2021) and/or production of nitrous oxide—a greenhouse gas (Reay et al., 2012). Application of fertilizer N in excess of the crop needs may

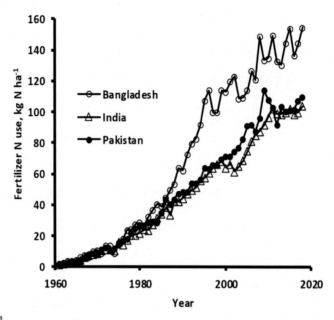

**FIGURE 7.1**

Average annual fertilizer nitrogen use per unit arable land in Bangladesh, India, and Pakistan from 1961 and 2018.

*Data source: (http://www.fao.org/faostat/en/#data/RFN, accessed 27 May 2021).*

also lead to increased residual inorganic N in the soil, which may accelerate the decomposition of soil organic matter and deteriorate soil health (Bijay-Singh, 2018). Thus, it is important to ensure high yields along with high NUE in crop production systems. The challenge ahead is to apply not only an adequate amount of N but also wisely manage it to ensure that its large portion is used by the crop and minimal amount escapes the farms (Davidson et al., 2015). If NUE is not increased substantially, fertilizer N consumption will have to be increased by about 60% to ensure global food security at the cost of major environmental issues (Ladha et al., 2016). By ensuring rapid improvement in NUE, farmers will be economically benefitted and there will be a reduction in N-based global warming and environmental pollution (Houlton et al., 2019).

In 2020, the Food and Agriculture Organization of the United Nations established an online database under the domain name *Soil Nutrient Budget* (http://www.fao.org/faostat/en/#data/ESB). It provides annual estimates of N inputs to the soil via fertilizers, organic manures, biological N fixation, and atmospheric deposition, and N removal through crop yields. The data are available from 1961 onwards and for different countries of the world. In this chapter, challenges and issues pertaining to NUE in South Asia have been discussed using these datasets. Since 99.08% of the total fertilizer N consumed in the South Asia during 2018 was used in Bangladesh, India, and Pakistan (http://www.fao.org/faostat/en/#data/RFN, accessed 27 May 2021), the datasets available at the *Soil Nutrient Budget* domain of the FAOSTAT provided an excellent opportunity to comparatively investigate different aspects of NUE in the three countries. The research being carried out to address issues for achieving high NUE in the region has also been discussed in the light of challenges being faced by farmers and policy makers.

## 7.2 Nitrogen use efficiency in relation to soil nitrogen

A substantial amount of soil N exists in the form of organic matter. As compared to the amount of N input through fertilizers, organic manures, biological N fixation, and atmospheric deposition, the soil organic N pool is several times bigger. For example, a typical soil in the Indo-Gangetic Plains of South Asia with total N content of 0.5 g kg$^{-1}$ soil (or 0.05%) in the field up to 0.3 m depth will contain about 2000 kg N ha$^{-1}$ in the soil N pool. Nitrogen from the soil N pool is continuously mineralized at rates determined by moisture and temperature and can become available to crop plants in the form of ammonium and nitrate ions. Even when N is applied through different sources including fertilizers, crop plants take more N from the soil N pool rather than from the applied N (Dourado-Neto et al., 2010; Gardner and Drinkwater, 2009; Sebilo et al., 2013). Ladha et al. (2005) reported a global average fertilizer N recovery of 44% (using $^{15}$N labeled fertilizers) by rice, wheat, and maize. In a cropping season, a major part of the applied N becomes a part of the big soil N pool; only a small portion is directly taken up by plants or is lost from the soil—pant system by different mechanisms. Applied N that becomes a part of the soil N pool will become available to crop plants in subsequent years.

Significance of N supplied by soil N pool to crop plants in defining NUE has important implications for both N nutrition of crops as well as environmental degradation due to surplus N. It can be very well understood from a typical example of growing wheat in a soil in the Indo-Gangetic plain of South Asia. If wheat is supplied with 120 kg N ha$^{-1}$ through fertilizer, 40% recovery efficiency of N (REN) is commonly observed in a well-managed crop. With a total N uptake by wheat to the tune of 110 kg N ha$^{-1}$, contribution of fertilizer N in it will be only 48 kg N ha$^{-1}$; 62 kg N ha$^{-1}$ will come from the soil N pool. In case, due to less soil organic C or inadequate management, the contribution of soil N pool to total N uptake by wheat is reduced from 62 to 50 kg N ha$^{-1}$, to achieve the same yield level by ensuring total N uptake of 110 kg N ha$^{-1}$, fertilizer N rate will have to be increased by 25% (to 150 kg N ha$^{-1}$), that too if fertilizer NUE remains 40%; REN normally decreases with increasing fertilizer N rate. Also at high N application levels, fertilizer N substitution value of soil N increases substantially. Thus, soil health in terms of soil organic matter levels is very crucial to ensure high yield levels. Bijay-Singh (2018) has reported that long-term use of high fertilizer N rates can be detrimental to soil health. By studying the global N budget in maize, rice, and wheat production systems, Ladha et al. (2016) reported that soil N reserves have been reduced by about 8% under maize or wheat; under rice cropping soil N increased by 4%. However, in South Asia where crop production has a long history, in most of the cropping systems size of the soil organic N pool has reached a steady-state or changing very slowly and N inputs from biological $N_2$ fixation and atmospheric deposition are relatively constant. Thus, overall improvement in NUE in crop production in South Asian countries can be achieved by ensuring high N uptake efficiency from applied fertilizer N and by not allowing increased use of N from soil organic N pool (Cassman et al., 2002).

## 7.3 Time trends of the total input, output, and surplus nitrogen in the soil and nitrogen use efficiency in South Asia

Data pertaining to annual average N input per unit area of cropland (kg N ha$^{-1}$ year$^{-1}$) through synthetic fertilizer (Nf), organic manures (Nom), biological N fixation (Nbf), atmospheric deposition (Ndep), and N removal in crop yield (Ny) from 1961 to 2018 were obtained for Bangladesh, India, Pakistan, and the World from the *Soil Nutrient Budget* domain of the FAOSTAT database (http://www.fao.org/faostat/en/#data/ESB, accessed 25 May 2021). Since units of both input and output of N are kg N ha$^{-1}$ year$^{-1}$, it is possible to use these data to estimate surplus N in the soil (Nsur) as the difference between total N input (Nt) and N output in the form of crop yield (Ny), and NUE as the N output expressed as the percentage of Nt as depicted through the following equations:

$$Nt = Nf + Nom + Nbf + Ndep \qquad (7.1)$$

$$NUE = 100 \times Ny \div Nt \qquad (7.2)$$

$$Nsur = Nt - Ny \qquad (7.3)$$

The time trends of Nt, Ny, NUE, and Nsur for the period 1961 to 2018 for Bangladesh, India, Pakistan, and the world are plotted in Fig. 7.2.

The total N input per unit of cropland area in India and Pakistan has been almost similar from 1961 to 2018 (Fig. 7.2). However, substantially more N has been applied to arable soils in Bangladesh than in India or Pakistan. In 2018, Nt in Bangladesh was 40% more than in India or Pakistan. Due to higher total N input in Bangladesh, the removal of N in crop yield (Ny) in 2018 was 1.93 and 2.72 times higher than in India and Pakistan, respectively. Interestingly, while Nt in India and Pakistan was comparable, Ny in India was higher than in Pakistan and was comparable with the world average. Lower Ny in Pakistan than in India could be due to climate and/or slightly better fertilizer and crop management practices in India.

During the 1960s, due to very small total N input, high NUE values were observed in all three countries (Fig. 7.2). However, the time trends in NUE were dictated by continuously increasing Nt and the varying rate of increase in N output

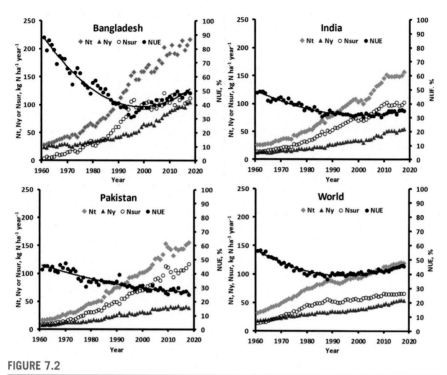

**FIGURE 7.2**

Time trend of total N input (Nt), N removal in crop yield (Ny), N use efficiency (NUE), and surplus N (Nsur) for Bangladesh, India, Pakistan, and the world for the period 1961 to 2018.

*Data source: (http://www.fao.org/faostat/en/#data/ESB, accessed 25 May 2021).*

since 1961. In Bangladesh, NUE declined from 88.1% in 1961 to 30.9% in 1995 but since then it has increased to 48.3% in 2018. In India, NUE of 47.7% in 1961 touched the lowest values around 30% in 1986 with a small improvement to about 35% by 2018 (Fig. 7.2). In case of Pakistan, NUE has been continuously falling from 45% to about 25% in 2018 thereby suggesting that fertilizer and crop management need to be improved urgently. Interestingly, global NUE values touched the lowest values around 1990 and since then these are exhibiting an increasing trend.

Surplus N as an index of potential environmental degradation due to reactive N and defined as the difference between Nt and Ny is continuously increasing in the three South Asian countries since 1961 (Fig. 7.2). As its magnitude is determined by both total N input and N output in the form of crop yield, the Nsur values differ considerably among Bangladesh, India, and Pakistan. Based on the average Nt and Ny values for 5 years during 2014−2018, Nsur as a percent of Nt was 52.4% in Bangladesh, 65.7% in India, and 73.5% in Pakistan. The highest values in Pakistan were due to low Ny than in India and Bangladesh.

## 7.4 Nitrogen output in the form of crop yield as a function of total N input

Nitrogen in harvested yield of crops generally responds to both organic and inorganic fertilizer N inputs following the Michaelis–Menten type of functional relationship and approaches a maximum yield. As total N input and N output in the yield are expressed in the same units, it should be possible to develop a relationship between them, which can help understand the course of NUE in the three South Asian countries. According to Lassaletta et al. (2014), the function relating Ny with Nt should obey three properties: (i) Ny should be zero at Nt = 0 because in the long run N removed in the harvest cannot exceed N restitutions to the soil, (ii) at very low Nt values, Ny should be equal to Nt, which means slope should be 1 because N-limited systems with low N supply are characterized by an NUE approaching 1, and (iii) Ny should plateau at high values of Nt as per the classical law of diminishing returns and also some other limiting factors may always impose a ceiling under very high and saturating N availability scenarios. A hyperbolic function with only one parameter obeyed these conditions (Lassaletta et al., 2014):

$$Ny = \frac{Nymax \times Nt}{Nymax + Nt} \tag{7.4}$$

where Nymax represents the maximum yield (as N) under saturating N input regimes. Eq. (7.4) also allows estimating Nt at which 0.5 Nymax is reached. The data pertaining to Nt and Ny for Bangladesh, India, and Pakistan were fitted to this function and values of Nymax were computed based on the complete period from 1961 to 2018 and for the last 15 years from 2004 to 2018. The actual data points, the best fit line and the Nymax values based on two time periods are shown in Fig. 7.3.

The Nymax values representing protein yield from cropland at maximum N input and estimated using data pertaining to the period from 1961 to 2018 were the highest for Bangladesh (122.7 kg N ha$^{-1}$ year$^{-1}$) and the lowest for Pakistan (44.3 kg N ha$^{-1}$ year$^{-1}$). The value for India was somewhat higher than Pakistan but less than half of the value for Bangladesh (Fig. 7.3). The magnitude of Nymax values indicates the efficiency with which applied N to cropping systems is converted to yield. The two Nymax values estimated from 58 years of data (1961–2018) and 15 years of data (2004–2018) characterize the yield potential under past and current N and crop management scenarios. The Nymax values of 122.7 and 161.7 kg N ha$^{-1}$ year$^{-1}$ for the two periods suggest that there is substantial improvement in the management of fertilizer N and overall crop management practices during the last 2 decades. It resulted in the production of high yield levels or high Ny even at N input levels even more than 200 kg N ha$^{-1}$ year$^{-1}$. In sharp contrast to Bangladesh, the Nymax value of 44.3 kg N ha$^{-1}$ year$^{-1}$ for Pakistan as estimated using the data for the period from 1961 to 2018, increased only by 9 kg N ha$^{-1}$ year$^{-1}$ when Nymax was estimated using data for the period from 2004 to 2018. It emphatically suggests that current agronomic and N management

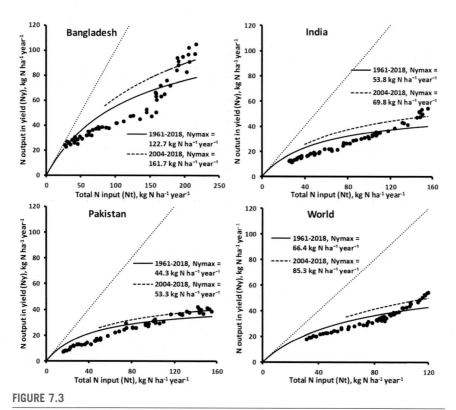

**FIGURE 7.3**

The N output in yield (Ny) versus total N input (Nt) plots, the best fit relationships between Ny and Nt as per Eq. (7.1) and Nymax values based on 1961–2018 and 2004–2018 periods for Bangladesh, India, Pakistan, and the world.

*Source: Recalculated, data from FAOSTAT 2021.*

practices in Pakistan are almost similar to those practiced more than 2 decades ago. The case for efficient management of N in India is somewhat better than Pakistan but inferior to Bangladesh.

Performance of agriculture in a region can be characterized in terms of yield gap defined as the difference between actual yield and the potential yield or the maximum yield which can be obtained in the absence of limitations of nutrients, water, and with efficient pest and disease control (van Ittersum et al., 2013). On similar lines, Lassaletta et al. (2014) proposed an N-based yield gap as (Nymax − Ny)/Nymax. It provides a relative difference between the maximum N output in the yield under saturating fertilizer N application regimes and the actually observed yield (Ny) at a given location and time. It should serve as a useful indicator of the degree of N limitation in a region or country. Nitrogen-based yield gaps for Bangladesh, India, and Pakistan for 2018 using Nymax computed on the basis of 2004−2018 data turned out to be 35.2% for Bangladesh, 22.6% for India, and 27.9% for Pakistan. According to Lassaletta et al. (2014), values of N-based yield gap less than 30%

indicate the limited possibility of enhancing yield levels by applying increasing levels of fertilizer N to crop production systems as well as low NUE. It suggests that both India and Pakistan need to increase NUE via technology upgradation to achieve high Nymax so that increased yield levels can be obtained at high fertilizer N levels.

## 7.5 Nitrogen use efficiency as a function of total and fertilizer N input

Overall trends in NUE for Bangladesh, India, and Pakistan as a function of Nt are shown in Fig. 7.4. When yields increase as a result of increasing N input along a fixed response function, decreasing trends in NUE with increasing N input are expected. As observed for all the three countries up to N input level of 150 kg N ha$^{-1}$, it represents the phase of agricultural expansion in which a rapid increase in fertilizer N use was observed but with a moderate increase in N output or yield (Lasaletta et al., 2014; Conant et al., 2013). Both India and Pakistan are still in this phase of agricultural intensification in which NUE is continuously falling with increasing N input into the cropping systems. However, Bangladesh seems to be entering into the next phase of agricultural intensification in which not only fertilizer N application is increasing but also the applied N is being utilized more efficiently by crops due to the adoption of better fertilizer N management practices than in India

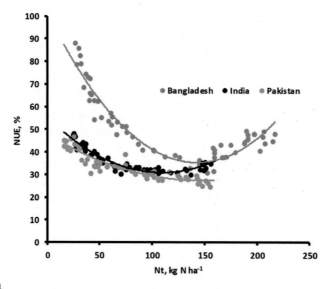

**FIGURE 7.4**

Nitrogen use efficiency expressed as the percentage of total N input (Nt) ending up in the crop yield as N output (Ny) as a function of Nt in Bangladesh, India, and Pakistan.

*Source: Recalculated, data from FAOSTAT 2021.*

and Pakistan. There are reports that farmers in large areas under rice have adopted fertilizer management strategies such as urea deep placement (UDP), which ensures significantly higher NUE and rice yields, and reduced losses of N (Nash et al., 2016). UDP has been anticipated to be adopted by farmers on 1.1 million ha of Aman rice and 700,000 ha of boro rice in Bangladesh. While the NUE trend in Fig. 7.4 for India shows that improvement in NUE has started but the trend of increasing NUE is still absent in the case of Pakistan. It is a matter of great concern because continuously falling NUE means a substantial portion of the applied N is being lost from the soil—plant system, which may pose threat to environmental security.

The proportion of fertilizer N in total N applied to crops in Bangladesh, India, and Pakistan is also linked with NUE. From 1961 to 2018, the proportion of Nf in Nt increased from 8% to 71% in Bangladesh and Pakistan and from 6% to 67% in the case of India. There exists a linear relationship between falling NUE and increasing proportion of fertilizer N in the total N applied (Fig. 7.5). Higher NUE in Bangladesh even with an increasing proportion of fertilizer N is conspicuously visible. Shahzad et al. (2019) has reported that among the three South Asian countries, the highest NUE and partial factor productivity of fertilizer N (PFPN) for both rice and wheat are the highest in Bangladesh and the lowest in Pakistan. Globally, the decline in PFPN and NUE was observed until 1980 but later on an increasing trend for rice or stabilizing trend for wheat were recorded. According to Shazad et al. (2019), N output in yield of wheat and rice in Pakistan showed little or no increase in response to increasing N input, primarily through synthetic fertilizers.

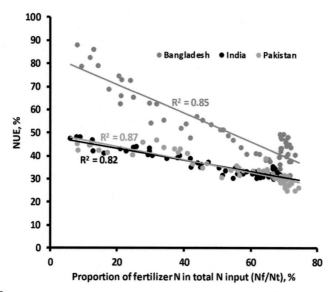

**FIGURE 7.5**

Nitrogen use efficiency in Bangladesh, India, and Pakistan as a function of proportion of fertilizer N in the total N applied.

*Source: Recalculated, data from FAOSTAT 2021.*

Thus, they reported a decline of NUE in wheat from 54% in 1961–65 to 34% in 2010–14. The corresponding decrease in NUE in rice was from 46% to 29%. In 2014/15, of the total fertilizer N consumption in India, 37% was in wheat, 46% in rice, and 8% in maize. In sharp contrast, 95% of total fertilizer N consumed in Bangladesh was used for rice production. In Pakistan, respectively, 67%, 22%, and 7% of the total fertilizer N consumed was used for wheat, rice, and maize (Heffer et al., 2017). According to an estimate by Ray et al. (2012), in India, stagnation of grain yields has been observed in 70%, 36%, and 31% land area under wheat, rice, and maize, respectively. On the other hand, yield improvement was observed in 100% land area under rice in Bangladesh, which explains the high NUE in Bangladesh. Apart from excessive use of fertilizer N in some regions, one of the important causes of low NUE in India and Pakistan is the imbalanced use of fertilizers (Ladha et al., 2020; Wakeel, 2015; Raza et al., 2018). In India, farmers have developed a tendency to rely primarily on N fertilizers to maximize crop yields. Not balancing high fertilizer N application levels with P and K results in negative effects on cereal yields and NUE (Bijay-Singh, 2017). Not following the fertilizer recommendations by farmers also leads to low NUE. According to Tahir et al. (2008), application of fertilizers by not following appropriate methods is common in Pakistan and it results in inefficient use of fertilizer N.

## 7.6 Nitrogen use efficiency and surplus nitrogen in the soil

In agricultural soils used for crop production, surplus N (Nsur) defined as the difference between total N input (Nt) and N output in the form of crop yield (Ny) gives an estimate of the potential N losses to the environment (van Beek et al., 2003; Van Groenigen et al., 2010). When Nt and Ny are expressed in the same units, Nsur as a robust measure of N losses from agricultural production systems is simple to calculate based on readily available data. It is easy to understand and provides a useful measure to demonstrate to the concerned public that agriculture can successfully control N losses while also improving the NUE in the cropping systems (McLellan et al., 2018). Because, Nt, Ny, and Nsur are expressed in the same units of kg N ha$^{-1}$ and NUE is expressed as Ny/Nt, Nsur is related with Ny and NUE as follows:

$$Nsur = Ny \times \left( \frac{1}{NUE} - 1 \right) \tag{7.5}$$

The concept of surplus N is valid at the farm, large watersheds (Thorburn and Wilkinson, 2013; Cela et al., 2017), and country levels (Zhang et al., 2015a). The relationship expressed through Eq. (7.5) shows that food security, environmental protection, and climate change mitigation associated with N$_2$O emissions revolve around increasing NUE in agricultural food production systems.

The surplus N is plotted against N output in Fig. 7.6. When the relationships between Nsur and Ny were fitted as per Eq. (7.5), the slope of the lines for Bangladesh, India, and Pakistan revealed the overall NUE for the three countries. As was

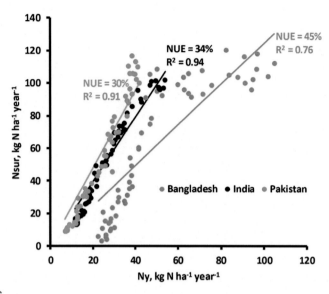

**FIGURE 7.6**

Surplus N (Nsur) as a function of N output in the form of grain yield (Ny) and N use efficiency (NUE) for crop production systems in Bangladesh, India, and Pakistan. The linear relationships and magnitudes of NUE are computed as per Eq. (7.5).

expected, Bangladesh showed the highest NUE value of 45%. Although India and Pakistan are fairly close with respect to NUE, the lowest value is revealed for Pakistan. The magnitude of overall NUE reveals that crop production in Pakistan is associated with the highest surplus N. In other words, high yield levels recorded in Pakistan are because of high N input levels but these are associated with huge environmental cost in terms of high Nsur values. In sharp contrast, farmers in Bangladesh can achieve high yield levels or Ny values but with relatively small amounts of surplus N. Performance of India in terms of producing surplus N from agricultural crop production is only slightly better than Pakistan.

In Fig. 7.7, surplus N in Bangladesh, India, and Pakistan is plotted against total N input into agricultural production systems. The 1:1 line shown in Fig. 7.7 depicts the hypothetical situation when NUE approaches zero, which means all the applied N leaves the soil—plant system as surplus N. Farther the relationship of a country from the 1:1 line toward the right, lesser the production of surplus N and it all depends upon NUE. Thus, agriculture in Bangladesh produced almost as much surplus N from total N input of 200 kg N ha$^{-1}$ year$^{-1}$ as is produced from application of 150 kg N ha$^{-1}$ in India and Pakistan.

Surplus N provides an estimate by which total N supply from all sources including fertilizers and manures exceeds crop needs. A modest amount of Nsur is an integral part of any agricultural production system but a large Nsur forms a pool of reactive N in the soil that acts as a potential source of pollution because a

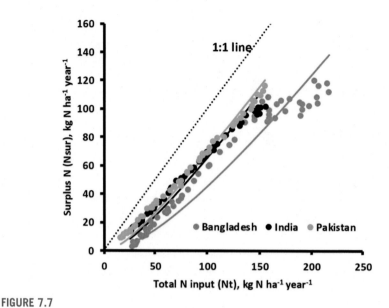

**FIGURE 7.7**

Surplus N (Nsur) as a function of total N input (Nt) into agricultural production systems in Bangladesh, India, and Pakistan.

*Source: Recalculated, data from FAOSTAT 2021.*

substantial portion of it can leave the soil—plant system via nitrate leaching, $N_2O$ emission, and ammonia volatilization. A portion of Nsur can also be used to maintain soil organic matter, but soils which are under crop production for several decades should reach a steady state with little or no change in soil organic N stocks (Ladha et al., 2020). Thus, Nsur in most of the soils in South Asian countries represents a robust estimate of the N pool in the soil that can be lost to the environment. McLellan et al. (2018) showed that Nsur is related to yield scaled $N_2O$ emission and nitrate leaching losses. The shape of the best-fit curves for the plots of N losses versus Nsur revealed that when total N input exceeds crop N needs, the surplus N becomes prone to loss, and that rate of N loss increases with increasing fertilizer N application rate. Recently, Sapkota et al. (2021) reported that in the Indo-Gangetic plains in India application of 18% less fertilizer N in the rice—wheat system than the farmer's fertilizer practice resulted in a 4%—12% increase in rice and wheat production and reduced $N_2O$ emission by about 2.5% in rice and between 12% and 20% in wheat. Shazad et al. (2019) reported that from 1961—2014, due to continuously declining NUE in Pakistan, N surplus increased linearly for cotton and wheat and in an exponential manner for rice. According to Raza et al. (2018), due to decline in NUE from 58% in 1961 to 23% in 2013, surplus N in Pakistan increased from 171 Gg N $year^{-1}$ to 3581 Gg N $year^{-1}$, which resulted in increased emissions of $NH_3$, $N_2O$, and NO from 70, 10, and 1 Gg N $year^{-1}$ to 1023, 155, and 46 Gg N $year^{-1}$, respectively. Recently, Varinderpal-Singh et al. (2021a) reported that by field-specific fertilizer N management using leaf color chart (LCC) and chlorophyll meter to guide need-based fertilizer N topdressings in baby corn, along with a 17% increase in yield and 52% increase in REN, average $NO_3$-N leachate load was reduced by 69% over blanket N use practice.

## 7.7 Challenges and options for enhancing nitrogen use efficiency

There are two major categories of strategies and methods to enhance NUE in agricultural production systems. In the first category, possibly the most effective strategy in South Asia is avoiding the application of fertilizer N in doses more than the need of the crops. Some of the practices which are not followed by farmers because of ignorance are also included in the first category and these are as follows: balanced application of fertilizer nutrients, appropriate time of application and placement of fertilizers in the soil, proper coordination of fertilizer and irrigation management, and following the fertilizer recommendations developed and improved from time to time for different crop cultivars in a region. The second category of strategies to improve NUE consists of techniques to modify sources of fertilizer N or management of fertilizer in such a way that Nymax or N output in the form of crop yields obtained under saturating N input regimes is raised to a level higher than the current one. As shown in Fig. 7.3, at high Nymax, the same yield of a crop can be obtained by applying a reduced fertilizer N dose leading to improvement in NUE. Advancement in technology front to raise Nymax level can be achieved through fertilizer source modification such as by imparting slow-release properties or by controlling nitrification and urea hydrolysis through the addition of nitrification and urease inhibitors to urea, managing fertilizer N following site-specific N management (SSNM) principles using gadgets like LCCs, chlorophyll meters such as SPAD meter, canopy reflectance sensors such as GreenSeeker optical sensor, or by using Nutrient Expert decision support system. UDP technology as practiced on a large area under rice cropping in Bangladesh can also lead to increased Nymax and high NUE.

In South Asia, fertilizers are managed as blanket recommendations formulated on the basis of responses of different cereal crops averaged over large geographic areas having similar climate and landforms. The blanket recommendations are sometimes also linked with soil tests such as soil organic carbon content. Jat et al. (2014) reviewed nutrient management in wheat in South Asia and concluded that it generally responded to application up to $120-150 \, kg \, N \, ha^{-1}$. Significant responses of rice are obtained in the range of $100-125 \, kg \, N \, ha^{-1}$ (Bijay-Singh and Singh, 2017). However, N fertilizers in South Asian countries are highly subsidized, farmers in some regions tend to apply fertilizer N more than the recommended doses. In such cases, yield does not increase but excess amount of fertilizer N results in reduced NUE, high surplus N, and economic loss to the farmers. In Pakistan, Maqsood et al. (2016) demonstrated that fertilizer N needs to be applied according to the crop demand and nutritional status of the soil. In India, Prasad et al. (2000) summarized the effect of applying fertilizer N at different rates to rice and wheat and concluded that at 40−60, 61−80, and 121−180 kg N ha$^{-1}$ for rice, REN values were 36%, 40%, and 31% and PFPN values were 84, 48, and 33 kg kg$^{-1}$ N, respectively. Corresponding REN values for wheat were 74%, 58%, and 52%, and PFPN values were 84, 50, and 31 kg kg$^{-1}$ N, respectively. According to Swarup (2002), REN in maize increased from 16.2% to 30.2%, when P was applied along with N and to 32% when both P and K were applied along with N. Similarly, 32.2% REN observed in wheat with the application of only N increased to 51.4% by applying P along with N, and to 64% with N, P, and K. Recently, based on a study conducted in five states in India, Panwar et al. (2019) reported that for fertilizer N application rates ranging from 100 to 150 kg N ha$^{-1}$ for both rice and wheat, an average yield of rice observed with only N increased by 31% by applying N and P and by 55% by applying N, P, and K. Average wheat yield observed with fertilizer N alone also increased by 31% and 57% by applying P and PK along with N. Increases in PFPN for rice were 31% and 55% for NP and NPK over N alone and that for wheat were 29% and 57% when along with N, P, and PK were applied. Katyal et al. (1987) applied $^{15}$N-labelled urea to wheat in a field experiment conducted in a coarse-textured soil and reported 0.8 t ha$^{-1}$ higher grain yield, 48% higher REN, and 62% less losses of N if urea was top-dressed after irrigation rather than common farmer's practice of applying it before an irrigation event.

Improvement in NUE by shifting the response curve up through advancement in technology front has already been reported by several workers in South Asian countries. Recently, Bijay-Singh et al. (2020) reviewed SSNM studies conducted in South Asia using gadgets like LCC, chlorophyll meter (particularly SPAD meter), and canopy reflectance sensors (particularly GreenSeeker optical sensor). Studies based on guiding fertilizer N application as guided by leaf color measured by SPAD meter showed a range in REN between 30% and 55% as compared to 20%−45% in the farmer's fields. Ranges in PFPN were 31.0−77.3 kg kg$^{-1}$ N in SPAD-based SSNM and 27.9−42.8 kg kg$^{-1}$ N with farmer's fertilizer practice (FFP). In LCC-based SSNM evaluated in 553 on-farm locations in South Asia, average fertilizer N rates under FFP and SSNM were 123 and 91 kg N ha$^{-1}$, respectively. While average yield under FFP and SSNM were 6.43 and 6.51 t ha$^{-1}$, the ranges in PFPN were 39−62 and 51−102 kg/kg, respectively. In another set of 38 experiments conducted in India, Pakistan, and Bangladesh, under LCC-based SSNM, average fertilizer N use and PFPN range were 126 kg N ha$^{-1}$ and 25−48 kg kg$^{-1}$ N, respectively. The corresponding values under FFP were 100 kg N ha$^{-1}$ and 36−66 kg kg$^{-1}$ N. But the average yield under LCC-SSNM treatment (5.19 t ha$^{-1}$) was significantly higher than that observed under FFP (4.53 t ha$^{-1}$) (Bijay-Singh et al., 2020). Substantial increase in different measures of NUE in wheat and maize have been reported due to following SSNM as compared to blanket recommendation (Kumar et al., 2021; Varinderpal-Singh et al., 2011, 2012, 2017; Umesh et al., 2018; Khurana et al., 2008; Bijay-Singh et al., 2018). Bijay-Singh et al. (2015) used GreenSeeker canopy reflectance sensor to practice SSNM on nine rice cultivars and observed a 10%—32.5% reduction in fertilizer N application but with no significant change in yield levels as compared to the blanket recommendation. As a result, the REN increased by 9.8%—41.4% by following sensor-based SSNM. Even in 19 on-farm comparisons, sensor-based SSNM plots needed 6.55%—51.8% less fertilizer N to produce up to 21.8% more yield than the FFP. Thus, PFPN increased in the range of 2%—159% by following SSNM. In wheat too, GreenSeeker optical sensor—based SSNM increased REN in the range of 12.3%—38.6% and 26.7%—37.5% less fertilizer N use in four cultivars than when fertilizer N was managed as per blanket recommendation (Bijay-Singh et al., 2017). Recently, Varinderpal-Singh et al. (2021b) reported a 34.1% increase in REN along with 25.8% less fertilizer N use and no significant change in wheat yield by managing fertilizer N following sensor-based SSNM rather than the general blanket recommendation.

A few studies in South Asia have revealed that controlled fertilizers and the use of urease and nitrification inhibitors can increase Nymax by pushing up the response curve of cereal crops to applied N. Patil et al. (2010) observed a 63% increase in PFPN in rice by applying a blend of polymer-coated urea and urea (50:50 N basis) over ordinary urea; it was accompanied by a yield increase of 15.6% and reduction in fertilizer N rate by 30%. Recently, in Pakistan, Perveen et al. (2021) reported a 16.0%—17.2% increase in NUE by applying polymer coated urea over ordinary urea to sunflower. Ghafoor et al. (2021) also recorded higher NUE with bioactive sulfur coated urea over ordinary urea in wheat. Katyal et al. (1987) reported an increase of 52.6% in REN in wheat by applying urea fortified with phenylphosphorodiamidate (PPD), a urease inhibitor, rather than ordinary urea. Several investigations have been carried out in South Asia to evaluate the nitrification inhibitor properties in the oil extracted from the seeds of neem (*Azadirachta indica* A. Juss). Bijay-Singh (2016) reviewed more than 75 such studies conducted in India and concluded that yield benefit in rice and wheat by applying neem oil coated urea rather than ordinary urea was between 5% and 6%. Mohanty et al. (2021) reported 6.2%—6.5% higher REN in transplanted rice due to the application of neem oil coated urea over ordinary urea. However, applying neem oil coated urea following SSNM principles resulted in an REN increase of 11.4%—14.6%. A 22.5% increase in REN by applying a single dose of polymer-coated urea than three splits of neem-coated urea in the Indian Punjab has been reported by Bhatt and Singh (2021).

UDP is widely accepted and effective fertilizer N management practice for lowland rice. It increases rice productivity and reduces the amount of fertilizer N to achieve optimum yield levels (Savant and Stangel, 1990; Bandaogo et al., 2015). In recent decades, UDP has been widely adopted by farmers in Bangladesh and resulted in a substantial increase in yield potential and NUE (Kabir et al., 2009; Hasanuzzaman et al., 2009; Islam et al., 2011; Miah et al., 2012; Sikder and Xiaoying, 2014; Rahman and Barmon, 2015). It has been reported that UDP increases NUE and rice grain yield by 50%—70% and 15%—20%, respectively; fertilizer N use is reduced by 30%—40% (Savant and Stangel, 1990; Alam et al., 2013; IFDC, 2013). After conducting 115 experiments in farmers' fields in Bangladesh, Majid Miah et al. (2016) reported that yield increase due to UDP over farmer's broadcast of prilled urea was 21%−31% higher during the Aus and Aman (wet) seasons and by 11%−17% in the Boro (dry) season. UDP also saved urea by 33% during the Aus and Aman seasons, and by 35% during the Boro season. The NUE estimated as PFPN ranged from 42 to 58 kg kg$^{-1}$ N with FFP and from 84 to 106 kg kg$^{-1}$ N with UDP. In Pakistan, Khalofah et al. (2021) has reported increased yield and NUE in basmati rice due to the placement of fertilizer N in the soil at a depth of 15 cm.

Different fertilizer N management scenarios being developed for improving NUE in crop production systems in South Asian countries will be adopted only if these have favorable cost:benefit ratios for the farmers. For example, farmers will adopt controlled-release N fertilizers or urea amended with urease and nitrification inhibitors only if the extra cost of the new products is less than the increased profits in crop yields. Large gains in NUE are also possible even by the widespread adoption of the existing technologies. Because complex socioeconomic factors affect decision-making by farmers for adopting new strategies, the first step in encouraging farmers to adopt new fertilizer management strategies is to make them adequately knowledgeable about NUE. According to Davidson et al. (2015), it is going to be critical in improving NUE. Analyzing economic and environmental aspects of emerging technologies, Zhang et al. (2015b) concluded that new fertilizer N management strategies which do not increase yield ceiling but result in low N application rates and reduced N losses do not find favor of farmers for adoption because these do not lead to land sparing. On the other hand, technologies, which increase the yield levels, are readily adopted by farmers due to high economic incentives. However, many times such technologies and management practices result in the application of high N rates and more N losses to the environment.

## 7.8 Conclusions

Total N input in an agricultural system and NUE determine both the crop production and the risk of global warming and environmental pollution linked with fertilizer N management at farm, region, or country level. Up to the mid-1980s, fertilizer N input per unit cropland area in Bangladesh, India, and Pakistan, the three largest countries in South Asia, was similar. By 2018, farms in Bangladesh were applying about 44% more fertilizer N per ha of land under crops than in India or Pakistan. Little or no improvement in maximum N output in the form of yield at saturating N input regimes (Nymax) even when based on N input−N output trajectories for 2004 to 2018 suggests that fertilizer N management in India and Pakistan did not change appreciably since the Green Revolution era in the 1960s. Bangladesh has not only resulted in increased N input in farming systems but also improved fertilizer N management strategies so that overall NUE in Bangladesh is 45%, as compared to 34% in India and 30% in Pakistan. High NUE values in Bangladesh in 1960s declined up to 1995 and then started increasing. The lowest NUE values in India were observed in 1986 and since then it is slowly increasing. However, in Pakistan, NUE is continuously falling since 1961. Nitrogen-based yield gaps for 2018 based on maximum N output in yield at saturating N input regimes calculated using 2004 to 2018 data are 35.2% for Bangladesh, 22.6% for India, and 27.9% for Pakistan. Values less than 30% indicate a limited possibility of improving yield levels by increasing fertilizer N application rates. Due to the high NUE observed in Bangladesh, it is generating substantially less surplus N per unit cropland area than India and Pakistan even though it is using higher fertilizer N application rates. With a long history of crop production, the agricultural lands in South Asia are already in a steady-state with respect to mineralization of soil organic matter, biological $N_2$ fixation, and atmospheric deposition, so that variations in fertilizer N application and management largely dictate crop production and surplus N, which can potentially leave the soil—plant system. The most effective strategy to improve NUE in South Asian countries is to avoid the application of fertilizer N more than the recommended dose in the regions where farmers have developed a tendency to apply excessive amounts of N. As SSNM ensures utilization of both fertilizer and soil N based on spatial and temporal variability in crop responsiveness to N, it seems to be emerging as an important strategy to improve NUE in crop production in South Asia. Particularly, LCC-based SSNM should be more attractive because it has a high cost:benefit ratio. Using chlorophyll meters or canopy reflective sensors to practice SSNM does not seem attractive to smallholder farmers in South Asia. Adoption of new sources such as controlled release N fertilizers or use of nitrification and urease inhibitors will depend upon how much their cost of production can be reduced by the fertilizer industry.

## References

Alam, M.M., Karim, M.R., Ladha, J.K., 2013. Integrating best management practice for rice with farmers' crop management techniques: a potential option for minimizing rice yield gap. Field Crop. Res. 144, 62–68. https://doi.org/10.1016/j.fcr.2013.01.010.

Bandaogo, A., Bidjokazo, F., Youl, S., Safo, E., Abaidoo, R., Andrews, O., 2015. Effect of fertilizer deep placement with urea supergranule on nitrogen use efficiency in Sourou Valley (Burkina Faso). Nutrient Cycl. Agroecosyst. 102, 79–89.

Bhatt, R., Singh, M., 2021. Comparative efficiency of polymer-coated urea for lowland rice in semi-arid tropics. Commun. Soil Sci. Plant Anal. https://doi.org/10.1080/00103624.2021.1925689.

Bijay-Singh, 2016. Agronomic Benefits of Neem Coated Urea—a Review. International Fertilizer Association, Paris. International Fertilizer Association Review Papers.

Bijay-Singh, 2017. Management and use efficiency of fertilizer nitrogen in production of cereals in India — issues and strategies. In: Abrol, Y.P., Adhya, T.K., Aneja, V.P., Raghuram, N., Pathak, H., Kulshrestha, U., Sharma, C., Bijay-Singh (Eds.), The Indian Nitrogen Assessment. Elsevier, pp. 149–162.

Bijay-Singh, 2018. Are nitrogen fertilizers deleterious to soil health? Agronomy 8, 48. https://doi.org/10.3390/agronomy8040048.

Bijay-Singh, Craswell, E., 2021. Fertilizers and nitrate pollution of surface and ground water: an increasingly pervasive global problem. SN Appl. Sci. 3, 1–24.

Bijay-Singh, Singh, V.K., 2017. Advances in nutrient management in rice cultivation. In: Sasaki, T. (Ed.), Achieving Sustainable Cultivation of Rice, vol. 2. Burleigh Dodds Science Publishing Limited, Cambridge, UK, pp. 25–68. https://doi.org/10.19103/AS.2016.0003.16.

Bijay-Singh, Varinderpal-Singh, Purba, J., Sharma, R.K., Jat, M.L., Yadvinder-Singh, Thind, H.S., Gupta, R.K., Choudhary, O.P., Chandna, P., Khurana, H.S., Kumar, A., Jagmohan-Singh, Uppal, H.S., Uppal, R.K., Vashistha, M., Gupta, R.K., 2015. Site-specific nitrogen management in irrigated transplanted rice (*Oryza sativa*) using an optical sensor. Precis. Agric. 16, 455–475. https://doi.org/10.1007/s11119-015-9389-6.

Bijay-Singh, Varinderpal-Singh, Yadvinder-Singh, Thind, H.S., Kumar, A., Choudhary, O.P., Gupta, R.K., Vashistha, M., 2017. Site-specific fertilizer nitrogen management using optical sensor in irrigated wheat in the North-Western India. Agric. Res. 6, 159–168. https://doi.org/10.1007/s40003-017-0251-0.

Bijay-Singh, Varinderpal-Singh, Yadvinder-Singh, Ajay-Kumar, Sharma, S., Thind, H.S., Choudhary, O.P., Vashistha, M., 2018. Site-specific fertilizer nitrogen management in irrigated wheat using chlorophyll meter (SPAD meter) in the North-Western India. J. Indian Soc. Soil Sci. 66, 53–65. https://doi.org/10.5958/0974-0228.2018.00006.3.

Bijay-Singh, Varinderpal-Singh, Ali, A.M., 2020. Site-specific fertilizer nitrogen management in cereals in South Asia. Sustain. Agric. Rev. 39, 137–178. https://doi.org/10.1007/978-3-030-38881-2_6.

Cassman, K.G., Dobermann, A., Walters, D.T., 2002. Agroecosystems, nitrogen-use efficiency and nitrogen management. AMBIO 31, 132–140. https://doi.org/10.1579/0044-7447-31.2.132.

Cela, S., Ketterings, Q.M., Soberon, M.A., Rasmussen, C.N., Czymmek, K.J., 2017. Upper Susquehanna watershed and New York State improvements in nitrogen and phosphorus mass balances of dairy farms. J. Soil Water Conserv. 72, 1–11.

Conant, R.T., Berdanier, A.B., Grace, P.R., 2013. Patterns and trends in nitrogen use and nitrogen recovery efficiency in world agriculture. Global Biogeochem. Cycles 27, 558–566. https://doi.org/10.1002/gbc.20053.

Davidson, E.A., Suddick, E.C., Rice, C.W., Prokopy, L.S., 2015. More food, low pollution (Mo Fo Lo Po): a grand challenge for the 21st century. J. Environ. Qual. 44, 305–311. https://doi.org/10.2134/jeq2015.02.0078.

Dourado-Neto, D., Powlson, D., Abu Bakar, R., Bacchi, O.O.S., Basanta, M.V., thi Cong, P., Keerthisinghe, G., Ismaili, M., Rahman, S.M., Reichardt, K., Safwat, M.S.A., Sangakkara, R., Timm, L.C., Wang, J.Y., Zagal, E., van Kessel, C., 2010. Multiseason recoveries of organic and inorganic nitrogen-15 in tropical cropping systems. Soil Sci. Soc. Am. J. 74, 139–152. https://doi.org/10.2136/sssaj2009.0192.

Gardner, J.B., Drinkwater, L.E., 2009. The fate of nitrogen in grain cropping systems: a meta-analysis of 15N field experiments. Ecol. Appl. 19, 2167–2184. https://doi.org/10.1890/08-1122.1.

Ghafoor, I., Habib-ur-Rahman, M., Ali, M., Afzal, M., Ahmed, W., Gaiser, T., Ghaffar, A., 2021. Slow-release nitrogen fertilizers enhance growth, yield, NUE in wheat crop and reduce nitrogen losses under an arid environment. Environ. Sci. Pollut. Res. 1–16. https://doi.org/10.1007/s11356-021-13700-4.

Hasanuzzaman, M., Nahar, K., Alam, M.M., Hossain, M.Z., Islam, M.R., 2009. Response of transplanted rice to different application methods of urea fertilizer. Int. J. Sustain. Agric. 1, 1–5.

Heffer, P., Gruère, A., Roberts, T., 2017. Assessment of Fertilizer Use by Crop at the Global Level 2014-2014/15. International Fertilizer Association (IFA), Paris, France, and International Plant Nutrition Institute (IPNI), p. 18. Report A/17/134 rev.

Houlton, B.Z., Almaraz, M., Aneja, V., Austin, A.T., Bai, E., Cassman, K.G., Compton, J.E., Davidson, E.A., Erisman, J.W., Galloway, J.N., Gu, B., 2019. A world of co-benefits: solving the global nitrogen challenge. Earth's Fut. 7, 865–872. https://doi.org/10.1029/2019EF001222.

IFDC, 2013. Fertilizer Deep Placement. International Fertilizer Development Center, Muscle Shoals, AL. IFDC Solutions. http://issuu.com/ifdcinfo/docs/fdp_8pg_final_web?e=1773260/1756718.

Islam, M.S., Rahman, F., Hossain, A.T.M.S., 2011. Effects of NPK briquettes on rice (*Oryza sativa*) in tidal flooded ecosystem. Agriculturists 9, 37–43.

Jat, M.L., Bijay-Singh, Gerard, B., 2014. Nutrient management and use efficiency in wheat systems of South Asia. Adv. Agron. 125, 171–259. https://doi.org/10.1016/B978-0-12-800137-0.00005-4.

Kabir, M.H., Sarkar, M.A.R., Chowdhury, A.K.M.S.H., 2009. Effect of urea super granules, prilled urea and poultry manure on the yield of transplant Aman rice varieties. J. Bangladesh Agric. Univ. 7, 259–263.

Katyal, J.C., Bijay-Singh, Vlek, P.L.G., Buresh, R.J., 1987. Efficient nitrogen use as affected by urea application and irrigation sequence. Soil Sci. Soc. Am. J. 51 (2), 366–370.

Khalofah, A., Khan, M.I., Arif, M., Hussain, A., Ullah, R., Irfan, M., Mahpara, S., Shah, R.U., Ansari, M.J., Kintl, A., Brtnicky, M., Danish, S., Datta, R., 2021. Deep placement of nitrogen fertilizer improves yield, nitrogen use efficiency and economic returns of transplanted fine rice. PLoS One 16, e0247529. https://doi.org/10.1371/journal.pone.0247529.

Khurana, H.S., Phillips, S.B., Bijay-Singh, Alley, M.M., Dobermann, A., Sidhu, A.S., Yadvinder-Singh, Peng, S., 2008. Agronomic and economic evaluation of site-specific nutrient management for irrigated wheat in Northwest India. Nutrient Cycl. Agroecosyst. 82, 15–31. https://doi.org/10.1007/s10705-008-9166-2.

Kumar, D., Patel, R.A., Ramani, V.P., Rathod, S.V., 2021. Evaluating precision nitrogen management practices in terms of yield, nitrogen use efficiency and nitrogen loss reduction in maize crop under Indian conditions. Int. J. Plant Prod. 15, 243–260.

Ladha, J.K., Pathak, H., Krupnik, T.J., Six, J., van Kessel, C., 2005. Efficiency of fertilizer nitrogen in cereal production: retrospects and prospects. Adv. Agron. 87, 85–156.

Ladha, J.K., Tirol-Padre, A., Reddy, C.K., Cassman, K.G., Verma, S., Powlson, D.S., Van Kessel, C., Richter, D.D.B., Chakraborty, D., Pathak, H., 2016. Global nitrogen budgets in cereals: a 50-year assessment for maize, rice and wheat production systems. Sci. Rep. 6, 1–9. https://doi.org/10.1038/srep19355.

Ladha, J.K., Jat, M.L., Stirling, C.M., Chakraborty, D., Pradhan, P., Krupnik, T.J., Sapkota, T.B., Pathak, H., Rana, D.S., Tesfaye, K., Gerard, B., 2020. Achieving the sustainable development goals in agriculture: the crucial role of nitrogen in cereal-based systems. Adv. Agron. 163, 39–116.

Lassaletta, L., Billen, G., Grizzetti, B., Anglade, J., Garnier, J., 2014. 50 year trends in nitrogen use efficiency of world cropping systems: the relationship between yield and nitrogen input to cropland. Environ. Res. Lett. 9, 105011. https://doi.org/10.1088/1748-9326/9/10/105011.

Maqsood, M.A., Awan, U.K., Aziz, T., Arshad, H., Ashraf, N., Ali, M., 2016. Nitrogen management in calcareous soils: problems and solutions Pakistan. J. Agric. Sci. 53, 79−95.

Mazid Miah, M.A., Gaihre, Y.K., Hunter, G., Singh, U., Hossain, S.A., 2016. Fertilizer deep placement increases rice production: evidence from farmers' fields in southern Bangladesh. Agron. J. 108, 805−812.

McLellan, E.L., Cassman, K.G., Eagle, A.J., Woodbury, P.B., Sela, S., Tonitto, C., Marjerison, R.D., van Es, H.M., 2018. The nitrogen balancing act: tracking the environmental performance of food production. Bioscience 68, 194−203.

Miah, I., Chowdhury, M.A.H., Sultana, R., Ahmed, I., Saha, B.K., 2012. Effects of prilled urea and urea super granule on growth, yield and quality of BRRI dhan28. J. Agroforest. Environ. 6, 57−62.

Mohanty, S., Nayak, A.K., Bhaduri, D., Swain, C.K., Kumar, A., Tripathi, R., Shahid, M., Behera, K.K., Pathak, H., 2021. Real-time application of neem-coated urea for enhancing N-use efficiency and minimizing the yield gap between aerobic direct-seeded and puddled transplanted rice. Field Crop. Res. 264, 108072.

Nash, J., Grewer, U., Bockel, L., Galford, G., Pirolli, G., White, J., 2016. Accelerating Agriculture Productivity Improvement in Bangladesh: Mitigation Co-benefits of Nutrient and Water Use Efficiency. CCAFS Info Note. Published by the International Center for Tropical Agriculture (CIAT) and the Food and Agriculture Organization of the United Nations (FAO).

Panwar, A.S., Shamim, M., Babu, S., Ravishankar, N., Prusty, A.K., Alam, N.M., Singh, D.K., Bindhu, J.S., Kaur, J., Dashora, L.N., Latheef Pasha, M.D., Chaterjee, S., Sanjay, M.T., Desai, L.J., 2019. Enhancement in productivity, nutrients use efficiency, and economics of rice-wheat cropping systems in India through farmer's participatory approach. Sustainability 11, 122. https://doi.org/10.3390/su11010122.

Patil, M.D., Das, B.S., Barak, E., Bhadoria, P.B., Polak, A., 2010. Performance of polymer-coated urea in transplanted rice: effect of mixing ratio and water input on nitrogen use efficiency. Paddy Water Environ. 8, 189−198. https://doi.org/10.1007/s10333-010-0197-3.

Perveen, S., Ahmad, S., Skalicky, M., Hussain, I., Habibur-Rahman, M., Ghaffar, A., Shafqat Bashir, M., Batool, M., Hassan, M.M., Brestic, M., Fahad, S., Sabagh, A.E.L., 2021. Assessing the potential of polymer coated urea and sulphur fertilization on growth, physiology, yield, oil contents and nitrogen use efficiency of sunflower crop under arid environment. Agronomy 11, 269. https://doi.org/10.3390/agronomy11020269.

Prasad, R., Singh, R.K., Rani, A., Singh, D.K., 2000. Partial factor productivity of nitrogen and its use efficiency in rice and wheat. Fert. News 45 (5), 63−65.

Rahman, S., Barmon, B.K., 2015. Productivity and efficiency impacts of urea deep placement technology in modern rice production: an empirical analysis from Bangladesh. J. Develop. Area. 49, 119−134.

Ray, D.K., Ramankutty, N., Mueller, N.D., West, P.C., Foley, J.A., 2012. Recent patterns of crop yield growth and stagnation. Nat. Commun. 3, 1−7.

Raza, S., Zhou, J., Aziz, T., Afzal, M.R., Ahmed, M., Javaid, S., Chen, Z., 2018. Piling up reactive nitrogen and declining nitrogen use efficiency in Pakistan: a challenge not challenged (1961−2013). Environ. Res. Lett. 13 (3), 034012.

Reay, D.S., Davidson, E.A., Smith, K.A., Smith, P., Melillo, J.M., Dentener, F., Crutzen, P.J., 2012. Global agriculture and nitrous oxide emissions. Nat. Clim. Change 2, 410−416. https://doi.org/10.1038/nclimate1458.

Sapkota, T.B., Jat, M.L., Rana, D.S., Khatri-Chhetri, A., Jat, H.S., Bijarniya, D., Sutaliya, J.M., Kumar, M., Singh, L.K., Jat, R.K., Kalvaniya, K., et al., 2021. Crop nutrient management using Nutrient Expert improves yield, increases farmers' income and reduces greenhouse gas emissions. Sci. Rep. 11, 1564. https://doi.org/10.1038/s41598-020-79883-x.

Savant, N.K., Stangel, P.J., 1990. Deep placement of urea supergranules in transplanted rice: principles and practices. Fert. Res. 25, 1–83.

Sebilo, M., Mayer, B., Nicolardot, B., Pinay, G., Mariotti, A., 2013. Long-term fate of nitrate fertilizer in agricultural soils. Proc. Natl. Acad. Sci. U. S. A 110, 18185–18189. https://doi.org/10.1073/pnas.1305372110.

Shahzad, A.N., Qureshi, M.K., Wakeel, A., Misselbrook, T., 2019. Crop production in Pakistan and low nitrogen use efficiencies. Nat. Sust. 2, 1106–1114.

Sikder, R., Xiaoying, J., 2014. Urea super granule (USG) as key conductor in agricultural productivity development in Bangladesh. Develop. Countr. Stud. 4, 132–139.

Swarup, A., 2002. Lessons from long-term experiments in improving fertilizer use efficiency and crop yields. Fert. News 47 (12), 9–12.

Tahir, M., Tanveer, A., Ali, A., Abbas, M., Wasaya, A., 2008. Comparative yield performance of different maize (*Zea mays* L.) hybrids under local conditions of Faisalabad-Pakistan. Pakistan J. Life Soc. Sci. 6, 118–120.

Thorburn, P.J., Wilkinson, S.N., 2013. Conceptual frameworks for estimating the water quality benefits of improved agricultural management practices in large catchments. Agric. Ecosyst. Environ. 180, 192–209.

Umesh, M.R., Swamy, T.S., Ananda, N., Shanwad, U.K., Chittapur, B.M., Desai, B.K., Angadi, S., 2018. Nitrogen application based on decision support tools to enhance productivity, nutrient-use efficiency and quality of sweet corn (*Zea mays*). Indian J. Agron. 63, 331–336.

van Beek, C., Brouwer, L., Oenema, O., 2003. The use of farmgate balances and soil surface balances as estimator for nitrogen leaching to surface water. Nutrient Cycl. Agroecosyst. 67, 233–244.

Van Groenigen, J., Velthof, G., Oenema, O., Van Groenigen, K., Van Kessel, C., 2010. Towards an agronomic assessment of $N_2O$ emissions: a case study for arable crops. Eur. J. Soil Sci. 61, 903–913. https://doi.org/10.1111/j.1365-2389.2009.01217.x.

van Ittersum, M.K., Cassman, K.G., Grassini, P., Wolf, J., Tittonell, P., Hochman, Z., 2013. Yield gap analysis with local to global relevance—a review. Field Crop. Res. 143, 4–17.

Varinderpal-Singh, Bijay-Singh, Yadvinder-Singh, Thind, H.S., Gobinder-Singh, Satwinderjit-Kaur, Kumar, A., Vashistha, M., 2012. Establishment of threshold leaf colour greenness for need-based fertilizer nitrogen management in irrigated wheat (*Triticum aestivum* L.) using leaf colour chart. Field Crop. Res. 130, 109–119. https://doi.org/10.1016/j.fcr.2012.02.005.

Varinderpal-Singh, Bijay-Singh, Yadvinder-Singh, Thind, H.S., Buttar, G.S., Satwinderjit-Kaur, Meharban-Singh, Sukhvir-Kaur, Bhowmik, A., 2017. Site-specific fertilizer nitrogen management for timely sown irrigated wheat (*Triticum aestivum* L. and *Triticum turgidum* L. spp. durum) genotypes. Nutrient Cycl. Agroecosyst. 109, 1–16. https://doi.org/10.1007/s10705-017-9860-z.

Varinderpal-Singh, Kaur, N., Kunal, Blestar-Singh, Kumar, J., Thapar, A., Ober, E.S., 2021a. Nitrate leaching from applied fertilizer is reduced by precision nitrogen management in baby corn cropping systems. Nutrient Cycl. Agroecosyst. 1–13. https://doi.org/10.1007/s10705-021-10156-3.

Varinderpal-Singh, Kunal, Gosal, S.K., Choudhary, R., Singh, R., Adholeya, A., 2021b. Improving nitrogen use efficiency using precision nitrogen management in wheat (*Triticum aestivum* L.). J. Plant Nutr. Soil Sci. 184, 371–377. https://doi.org/10.1002/jpln.202000371.

Varinderpal-Singh, Yadvinder-Singh, Bijay-Singh, Thind, H.S., Kumar, A., Vashistha, M., 2011. Calibrating the leaf colour chart for need based fertilizer nitrogen management in different maize (*Zea mays* L.) genotypes. Field Crop. Res. 120, 276–282. https://doi.org/10.1016/j. fcr.2010.10.014.

Wakeel, A., 2015. Balanced Use of Fertilizers in Pakistan: Status and Perspectives. International Potash Institute, Bern, Switzerland.

Zhang, X., Davidson, E.A., Mauzerall, D.L., Searchinger, T.D., Dumas, P., Shen, Y., 2015a. Managing nitrogen for sustainable development. Nature 528, 51–59. https://doi.org/10.1038/nature15743.

Zhang, X., Mauzerall, D.L., Davidson, E.A., Kanter, D.R., Cai, R., 2015b. The economic and environmental consequences of implementing nitrogen-efficient technologies and management practices in agriculture. J. Environ. Qual. 44, 312–324. https://doi.org/10.2134/jeq2014.03.0129.

# Mitigation and actions toward nitrogen losses in Pakistan

8

**Muhammad Sanaullah[1], Ahmad Mujtaba[1], Ghulam Haider[2], Hafeez ur Rehman[3], Fathia Mubeen[4]**

[1]*Institute of Soil and Environmental Sciences, University of Agriculture Faisalabad, Punjab, Pakistan;* [2]*Department of Plant Biotechnology, Atta-ur-Rahman School of Applied Biosciences, NUST, Islamabad, Pakistan;* [3]*Department of Agronomy, University of Agriculture Faisalabad, Punjab, Pakistan;* [4]*Soil and Environmental Biotechnology Division, National Institute for Biotechnology and Genetic Engineering, Faisalabad, Pakistan*

## 8.1 Introduction

Nitrogen plays a fundamental role in crop and livestock production, being the structural part of chlorophyll, DNA, RNA, and proteins. The global use of synthetic nitrogenous fertilizers has increased about eightfold since the industrial era, 1961 (Fowler et al., 2013b), and will increase up to 118.2 million tons with increasing population (10.5 billion in 2050) and its food demand (Sharma and Bali, 2018; Haroon et al., 2019).

A significant portion of the applied N is not available to crop plants due to leaching, denitrification, erosion, fixation, precipitation, runoff, or volatilization. In this way, annual input of N from chemical fertilizers and manure sources is inadequately (30−80%) utilized in most crop production and livestock production systems. Resultantly, the surplus N moves off crop fields and contaminates surface and ground water resources, and the atmosphere (Chapter 3). Excessive N application and low nitrogen use efficiency (NUE) are the main reasons responsible for economic losses as well as environmental pollution worldwide, including Pakistan (Raza et al., 2018). This excessive N use may have damaging effects on human health, especially in the regions where intensive agriculture and animal husbandry is practiced (Butterbach-Bahl et al., 2011). Nonjudicious use of chemical fertilizers in Pakistan's agriculture is deteriorating natural resources and resulting in suboptimal crops yields and low profit margin for famers (Wakeel, 2015). Pakistan is among the top four countries which produce important crops like wheat, cotton, and rice in terms of N-based fertilizer utilization. However, only 25−50% of this added N is utilized by the plants and the remaining is lost to the environment (Shahzad et al., 2019).

There is a dire need to increase NUE in agriculture for economic and environmental reasons and to devise and implement mitigation strategies to reduce N

Nitrogen Assessment. https://doi.org/10.1016/B978-0-12-824417-3.00001-0

pollution, especially for those countries having low NUE, including Pakistan. Various traditional practices alongside advance tools are currently being implemented to improve NUE of crops. As soil N status varies greatly, N management in crops depends upon rapid and appropriate application technologies along with the precised estimation of the available soil N, and seasonal crop N status.

Integrated N management (INM) may contribute in achieving most of the Sustainable Development Goals (SDGs), especially no poverty (SDG 1), zero hunger (SDG 2), good health and well-being (SDG 3), clean water and sanitation (SDG 6), climate action (SDG 13), life below water (SDG 14), and life on land (SDG 15). Currently, N pollution is adversely affecting air, climate, land, and water (Galloway et al., 2008; Bouwman et al., 2013; Fowler et al., 2013a; Sutton et al., 2019) which requires further action to improve NUE in agriculture. Thus, INM in agriculture has a dual benefit of decreased N losses and enhanced food production through balanced fertilization and circular economy principles.

In this chapter, the challenges and opportunities for N losses mitigation, including mitigations options for ammonia ($NH_3$), nitrate ($NO_3$), and nitrous oxides ($N_2O$), have been discussed and different action plans have been reviewed considering presently in hand and upcoming technologies to enhance N management, with special reference to Pakistan.

## 8.2 Nitrogen footprints: global versus regional realities of N losses mitigations

Recent research attention has been focused on the extent and impact of changes, but only few focuses on the use of resource entities that cause a large amount of reactive nitrogen (Nr) production. It is estimated that $\sim$50% of the world's population is now alive because of the increased supply of N fertilizer, elucidating the massive role of N has in fulfilling human food requirements (Erisman et al., 2008; Sutton et al., 2013). It is predicted that N needs will further increase in near future along with increasing human population (Godfray et al., 2010). Anthropogenic activities like land-use changes, urbanization, injudicious use of inorganic N fertilizers, and the globalization of food systems have resulted in altering natural N cycle (Vitousek et al., 1997; Fowler et al., 2013a).

One evaluation strategy is the use of N footprint (NF) which is the total Nr released into the environment due to an object's production to consumption. The NF research has theoretical as well as practical importance for evaluation of the human activities affecting the Nr emissions, for regulating human livelihood and reducing anthropogenic emissions: however, research on NF is still limited (Qin et al., 2011; Čuček et al., 2012). NF is a unique approach for quantifying environmental losses of Nr species resulting from human actions (Gu et al., 2013). The average of an individual NF can be estimated by dividing the number of the newly created Nr by the world's total population (Galloway et al., 2014). It indicates how an individual's diet consumption patterns affect N pollution (Oita et al., 2018).

The major NF tools are (1) N-calculator, used for individuals to estimate their contribution to Nr losses to the environment through their food consumption, transportation, electricity use, purchase of goods, and use of services, and (2) N-neutrality describes a way to compensate the NF that could not be reduced by any of the mitigation measures (Galloway et al., 2014). The food production NF is calculated with virtual N factors (VNFs), which describe the total N lost to the environment during production per unit of N in the final consumed food product. These food production—related N losses include fertilizer not utilized by the plant, crop residues, feed not incorporated into the animal products, processing waste, and household food waste. Recycling within the food production process (e.g., crop residue and manure recycled as fertilizer) is also accounted for in the VNF calculation (Leach et al., 2012). Average loss of Nr per inhabitant globally as well as of Asia is given in Fig. 8.1.

Whereas a country's NF has been defined as the production of commodities consumed in the country, and the amount of Nr emitted during consumption and transportation, regardless of whether these commodities are produced domestically or internationally. The per capita NF ranges from below 7 kg N $yr^{-1}$ in developing countries such as Liberia, Papua New Guinea, and Côte d'Ivoire to above 100 kg N $yr^{-1}$ in developed countries like Luxembourg and Hong Kong (Liu et al., 2016). China, India, Brazil, and the United States account for 46% of the global N emissions. About a quarter of the global NF comes from commodities that cross borders through trade. If we consider agricultural crop production and food trade, major countries contribute about 60% to the global N, i.e., China 22.2%, India 15.5%, United States 15.4%, Brazil 4.9%, Pakistan 2.5%, and Bangladesh 1.8%

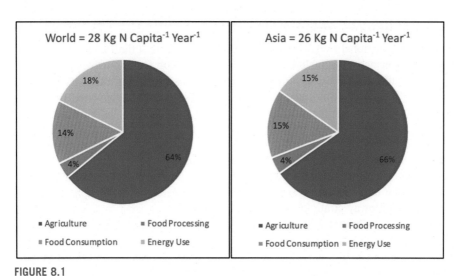

**FIGURE 8.1**

Average loss of Nr per inhabitant Global as well as Asia.

*Based on Galloway et al. (2014).*

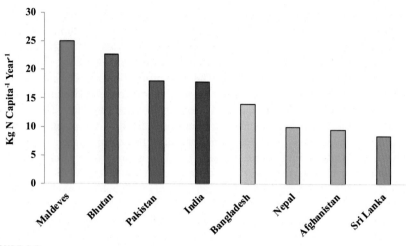

**FIGURE 8.2**

Per-capita ranking of South Asian countries.

*Adapted from Oita et al. (2016).*

(Liu et al., 2016). According to Fig. 8.2, per-capita ranking of NFs in kg per year for South Asian countries varies from 8.4 in Sri Lanka, 9.5 for Afghanistan, 10 for Nepal, 14 for Bangladesh, 17.8 for India, 18 for Pakistan, 22.6 for Bhutan, and 25 for Maldives (Oita et al., 2016).

## 8.3 Challenges and opportunities for mitigating N losses

The changes in global N cycle started from the second-half of the 19th century due to anthropogenic activities including rapid industrialization and intensive agricultural practices with heavy N use as fertilizer. Such activities are driving excessive releases of $CO_2$, $CH_4$ and Nr (Stark and Richards, 2008) because N is converted from one form to another through biogeochemical processes (Smil, 2004; Hatfield and Follett, 2008; Schlesinger and Bernhardt, 2013). Excluding $N_2$, compounds like $NH_3$, $NO_3$, nitrous oxide ($N_2O$), oxides of nitrogen ($NO_x$), amines, and organic forms are considered as reactive nitrogen (Nr) (Vitousek et al., 2013; Liu et al., 2019). Anthropogenic Nr contributes to smog, atmospheric haze, atmospheric acidification, reductions in biodiversity, $NO_3$ pollution in groundwater, and stratospheric ozone depletion (Galloway et al., 2003; Schlesinger, 2009).

The basic challenge in mitigation of N losses is to optimize NUE (Venterea et al., 2012) as shown in Fig. 8.3. Urea is extensively used as an N source along with sub-optimal P and little or 0 K, which is among the main reasons of low NUE (Zaman et al., 2008). Urea accounts for about 85% of overall N used in Pakistan during the years 1981−2013 (Heffer and Prud'homme, 2016), but its efficiency is low when compared with other nitrogenous fertilizers (Furtado da Silva et al., 2020).

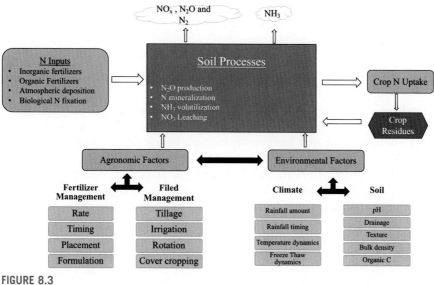

**FIGURE 8.3**

Schematic diagram of N losses and management.

*Adopted from Venterea et al. (2012).*

Once urea is applied to farmland, especially in calcareous and alkaline soils (like most of Pakistani soil), it is readily lost as $NH_3$ volatilization. Field trials have shown that grain yield of maize is the highest with calcium ammonium nitrate application, followed by ammonium sulfate as compared to urea fertilizer (Abbasi et al., 2013). Substantial research work has been done to investigate soil N emissions during the past few decades, but mitigation measures are still at the forefront of research as reliable practices have not been proven for various sites, crop systems, and growing seasons (Stark and Richards, 2008; Liu et al., 2019). So, integrated approaches are required for N management by avoiding trade-offs and allowing multiple benefits to society and the environment (Oenema et al., 2011).

Farmers have not yet adopted proper management practices to improve NUE due to lack of technical expertize, training, and socioeconomic and policy aspects. Mitigation of N emissions especially $N_2O$ from fertilized fields is principally challenging because of the complexity and the multiple processes involved in Nr production within soil in response to management practices and climatic events. Nonetheless, NUE can be improved by reducing or reusing biomass streams in the form of food waste, manure, and sewage. Consumption patterns of energy and food by the society also play a vital role in this case (Bellarby et al., 2013; Westhoek et al., 2014). In the recent past, legislations have been introduced for N losses mitigation from agricultural sources (Stark and Richards, 2008).

Over the past 40 years, agro-environmental research has focused on improving N cycling processes and strategies to reduce N losses to the environment. Some of the researches and mitigation strategies that can be used to reduce environmental N pollution (Stark and Richards, 2008; Shibata et al., 2017) have been discussed in this chapter.

## 8.4 Options for $NH_3$ mitigation

Agriculture accounts for about 50% of total global $NH_3$ volatilization (Sommer et al., 2004). These $NH_3$ emissions significantly contribute to air and water pollution (smog, eutrophication, etc.), biodiversity loss, and human health problems (Ti et al., 2019). However, the reaction mechanism of $NH_3$ emission from an applied fertilizer may differ depending on N fertilizer type and soil properties (Fenn and Hossner, 1985). Both N fertilizer and soil characteristics play an important role in $NH_3$ formation and escape from the system (Zia et al., 1992). For instance, the main driving force of $NH_3$ formation is soil pH. The $NH_4^+$ fertilizer leads to $NH_3$ loss when applied in high pH soils (Gavrilova et al., 2019). The soils in Pakistan are predominantly alkaline which favorably supports $NH_3$ emissions. In addition, the agricultural soils of Pakistan fall under both extremely warm arid and at some parts of the country it is temperate soil with extremely low temperature in winter months providing favorable conditions for $NH_3$ emission.

### 8.4.1 Choice of N fertilizers

All $NH_4^+$ containing nitrogenous fertilizers such as urea, an extensively used N fertilizer in Pakistan, are prone to $NH_3$ volatilization losses. As Pakistani soils possess alkaline soil pH and high temperature, understanding the choice of N fertilizer according to soil chemical properties, ecological conditions (atmospheric temperature) of farm and plant N uptake mechanisms are inevitable to reduce $NH_3$ losses and to improve NUE. So alternate to urea, fertilizers containing $NH_4NO_3$ or $NH_4SO_4$ could be used as these fertilizers have low potential risk of $NH_3$ volatilization (Rogers et al., 2018). In addition, various controlled release or slow-release fertilizers (polymer- or neem-coated urea) can be used which have shown 35−40% reduction in $NH_3$ losses (Rogers et al., 2018).

### 8.4.2 Fertilizer placement approaches

Different practices of fertilizer placement and land management may serve as one of the most cost-effective options for reducing $NH_3$ volatilization (Rogers et al., 2018). The broadcast application of urea and other $NH_4^+$-based fertilizers is a common practice in Pakistan which may cause high risk of $NH_3$ losses (Alia et al., 2016). In contrast, incorporation and subsurface banding will help to reduce $NH_3$ losses. The injection method may help to reduce $NH_3$ losses significantly, thereby, potentially increasing crop production and farm profit (Pan et al., 2016). Maximum reduction in $NH_3$ volatilization was achieved when urea and other $NH_4^+$-containing fertilizers were applied in a subsurface band at least two inches below the surface in alkaline calcareous soils (Jones et al., 2013). This approach is well researched, and a reduction of about 10–40% in NH$_3$ emission is reported. Moreover, 4Rs stewardship of FAO clearly advocates the importance of fertilizer placement in improving the use efficiency.

Recommended N application rates and placement methods should also be rationalized by considering seasonal soil available N (NO$_3^-$ and NH$_4^+$) and other soil properties such as pH, exchange properties of soil, temperature, etc. Therefore, the 4R nutrient stewardship concept of right fertilizer from right source, at right rate, right placement and right time (Bruulsema et al., 2009), may help in achieving reduced NH$_3$ volatilization losses. The primary management goal is to move urea and NH$_4^+$-fertilizers into the soil profile so they can be taken up by the crop, while minimizing NH$_3$ losses. As different crops have different N requirements (Table 8.1), N fertilizers should be applied as per requirement, incorporated within soil and should be added in split application to avoid NH$_3$ emissions.

### 8.4.3 Urease inhibitors

Urea once added to soil goes under hydrolysis with the help of an extracellular enzyme urease which is ubiquitous in the soils. Blocking the activity of urease enzyme will definitely affect the production of NH$_4^+$, a substrate for NH$_3$ emission. Certain chemicals are reported to reduce the urease activity like phosphoramidates, quinones, hydroquinone, benzothiazoles, thioureas, coumarin, and vanadium-hydrazine together with copper, zinc, sulfur, ammonium thiosulphate, oxidized charcoal, or silver nanoparticles, hence can be used as urease inhibitors (Modolo et al., 2018). A number of reports are available on beneficials impacts of using these inhibitors on improved NUE and crop growth (Engel et al., 2011; Rogers et al., 2018).

### 8.4.4 Livestock and poultry feeding

For livestock, reducing crude protein (CP) from ruminant diets (<15% of dry matter) could be an effective strategy for lowing NH$_3$ losses (Broderick, 2003; Swensson, 2003). While in poultry industry, controlling N excretion through feeding measures is very limited as compared with livestock because the conversion efficiency is already high and the variability within a flock of birds is greater. Nitrogen losses may be highest for grass-only or grass-legume summer feedstock by grazing of young, intensively fertilized grass or grass legume mixtures. Urine excreted by grazing animals typically infiltrates into the soil. This means that NH$_3$ emissions per animal are reduced by extending the periods in which animal graze compared with the time spent with animals housed, where the excreta are collected, stored, and applied to land. However, given the clear and well quantified effect on NH$_3$ emissions, increasing the period that animals are grazing all day can be considered as a strategy to reduce emissions (Reis et al., 2015).

**Table 8.1** An overview of recommended N application rates (kg ha$^{-1}$) in major crops in Punjab–Pakistan (year 2019–20).

| Rice | | Wheat | | Maize | | Sugarcane | | Cotton | |
|---|---|---|---|---|---|---|---|---|---|
| Fertile soil | Low OM soil ‡ | Fertile soil | Low-OM soil | Fertile soil | Low OM soil | Fertile soil | Low OM soil | Fertile soil | Low OM soil |
| †189.06 (coarse) 150.7 (fine) | – | 126.04 (irrigated) 93.16 (rain-fed) | 175.36 (irrigated) – | 205.5 (hybrid) 191.8 (normal) 93.16–126.04 (rain-fed) | 326.06 (hybrid) 252.08 (normal) | 238.38 | 328.8 | 219.2 (Bt-C-zone) 191.8 (Bt-Sec-zone) | 274 (Bt-C-zone) 246.6 (Bt-Sec-zone) |

*Fertile soil = when having organic matter (OM) = 1.29%, P = 14 ppm, K = 180 ppm; †Sowing after wheat crop; ‡ Low organic matter soil (OM = 0.86%, P = 7 ppm, K = 80 ppm).*

### 8.4.5 Livestock housing

Livestock housing could be a source of NH$_3$ losses. Following management strategies can help in reducing NH$_3$ losses.

#### 8.4.5.1 Segregation of urine and feces

A physical separation of urine and feces which contain *urease* in the housing system may minimize urea hydrolysis and NH$_3$ emissions from both housing and manure spreading (B. Moller et al., 2007; Burton, 2007; Fangueiro et al., 2008a; Fangueiro et al., 2008b). In addition, acidification and/or alkalization of segregated urine may inhibit urea hydrolysis. This separation of solid—liquid separation can have additional benefits during soil application, where urine infiltrates in soil more easily due to its lower dry-matter content than slurry, reducing NH$_3$ volatilization.

#### 8.4.5.2 Cleaning of floors in cattle houses

The grooved floor system along with "toothed" scrapers for livestock housing may be a reliable technique to decrease NH$_3$ emissions. These grooves should be equipped with perforations to allow drainage of urine. This results in a cleaner, low-emission floor surface with good traction for cattle to prevent slipping. Ammonia emission reduction ranges from 25% to 46% relative to the reference system (Swierstra et al., 2001). Similarly, cleaning of walking areas has the potential to substantially reduce NH$_3$ emissions.

#### 8.4.5.3 Frequent slurry removal

Frequent removal of liquid slurry to an outside store can substantially reduce NH$_3$ emissions by reducing the emitting surface and the slurry storage temperature.

#### 8.4.5.4 Bedding material

Bedding material in animal housing is an important factor for NH$_3$, N$_2$O, NO$_x$, and N$_2$ emissions. Its physical characteristics like urine absorbance capacity, bulk density, etc., are of more importance than chemical properties such as pH, CEC, C/N ratio, etc., in determining NH$_3$ emissions from dairy barn floors (Misselbrook and Powell, 2005; Powell et al., 2008; Gilhespy et al., 2009).

### 8.4.6 Poultry housing

Designs to reduce NH$_3$ emissions from poultry housing systems have been described in detail by Santonja et al. (2017), and are briefly described below.

#### 8.4.6.1 Rapid drying of poultry litter

For reducing NH$_3$ emissions, rapid drying of poultry manure can be achieved by ventilating the manure pit. The collection of manure on belts and the subsequent removal of manure to covered storage outside the building can also reduce NH$_3$ emissions, particularly if the manure has been dried on the belts through forced ventilation. The poultry manure dried up to 60—70% of the dry matter can minimize the subsequent formation of NH$_3$.

### 8.4.6.2 Use of air scrubbers

Use of acid scrubbers to treat exhaust air has been successfully employed in several countries (Melse and Ogink, 2005; Patterson, 2005; Ritz et al., 2006; Melse et al., 2012). Similarly, treatment of exhaust air by use of biotrickling filters (biological air scrubbers) has been successfully employed in several countries (Melse and Ogink, 2005; Patterson, 2005; Ritz et al., 2006; Melse et al., 2012). Biological scrubbers are also used to reduce $NH_3$ emissions by 70% (Ogink and Bosma, 2007; Melse et al., 2008).

## 8.4.7 Manure storage and treatment

In order to minimize $NH_3$ losses from animal and poultry manures, following measures should be adopted.

### 8.4.7.1 Covered storage of solid manure and slurry

For minimizing $NH_3$ emissions from manure and slurry, solid lids of metal or concrete, floating covers on lagoons and use of slurry bags could be used for the storage purposes (Bittman et al., 2014). A natural crust may form of slurry during its storage, which may substantially minimize $NH_3$ emissions (Bittman et al., 2014). Similarly organic manure can be covered with peat, clay, zeolite, phosphogypsum to prevent contact of $NH_3$ emitting surfaces with the air, especially when covering them with ammonium absorbing substances (Lukin et al., 2014).

### 8.4.7.2 Storage of solid manure under dry conditions

Animal and poultry manure can be stored in a dry place to reduce N emissions from a range of $N_r$ compounds and $N_2$. This is even more important for dried poultry litter, where keeping manure dry and out of the rain helps to avoid hydrolysis of uric acid to form $NH_3$ (Elliott and Collins, 1982).

### 8.4.7.3 Adsorption of N from slurry

Mineral additives such as clay/zeolite can adsorb $NH_4-N$ from the slurry and can thus potentially reduce $NH_3$ emissions. However, this can only be achieved effectively with high amount of additives, e.g., 25 kg of Zeolite per $m^3$ slurry have been shown necessary to adsorb 55% of $NH_4-N$ (Kocatürk-Schumacher et al., 2017, 2019). Similarly, struvite ($MgNH_4PO_4 \cdot 6H_2O$) can be precipitated for removal and recovery of N from slurry (Jensen, 2013). However, it only works for the N already present as $NH_4^+$ and further development is needed for appropriate application to liquid manures and digestates.

### 8.4.7.4 Slurry acidification

Another method to minimize $NH_3$ emissions from slurry is to reduce pH of the slurry by adding strong acids. In addition, slurry with a sufficiently reduced pH will also emit less $CH_4$, by up to 67%–87% (Petersen et al., 2012). Typically, $H_2SO_4$ is used for slurry acidification as it is the cheapest industrial acid and sulfate added through this acid can serve as plant nutrient source (Fangueiro et al., 2015).

### 8.4.8 Introduction of agroforestry

Agroforestry land uses include the cultivation of crops and trees with alternate rows of trees and annual crops, or block of trees in the landscape. This approach offers the opportunity for including unfertilized crops into the landscape, such as short-rotation coppices for bioenergy production. This can increase biodiversity, remove surplus N$_r$ from neighboring arable fields, minimize erosion, provide wind shelter, and can also increase deposition of NH$_3$ as surface roughness is increased (Sutton et al., 2004; Lawson et al., 2020). All these effects mitigate N$_r$ transportation at spatial scales and N$_r$ pollution of air and water (Pavlidis and Tsihrintzis, 2018).

## 8.5 Options for NO$_3$ leaching mitigation

Nitrate, a mobile anion, is highly soluble and can move in soil environment without being absorbed to the soil solids. An access of N in the environment provides potential opportunities to NO$_3$ leaching and threaten ground water quality (Jahangir et al., 2012). For instance, grassland ecosystems have been found as the hotspots of the NO$_3$ leaching due to animal excreta, applied slurries (Davies, 2000; Foster et al., 2000), and higher rates of synthetic N fertilizers applications (Vogeler et al., 2020). Therefore, intensive cropping systems with high N application are the potential sources of NO$_3^-$ leaching (Biernat et al., 2020).

Several products have been developed and made available in the market around the world for reducing/slowing down the N transformation process to synchronize the available N (NO$_3^-$ and NH$_4^+$) production and plant demands/uptake. Several natural allelopathic potential of crops and grasses (Subbarao et al., 2009; Haider et al., 2015) and chemical compounds have been identified and used to reduce the activity of microbes involved in nitrification process.

### 8.5.1 Nitrification inhibitors

Nitrification inhibitors (NIs) are the chemical compounds which can temporarily reduce the activities of nitrifying bacteria (*Nitrosomonas* and *Nitrobacter* bacteria) and are used to reduce NO$_3$ leaching from agricultural soils. The *Nitrosomonas* are responsible for conversion of NH$_4^+$ to nitrite (NO$_2^-$) and *Nitrobacter* to convert NO$_2^-$ into NO$_3^-$ in the soil (McKervey et al., 2005; Di et al., 2007). Thus, controlling the fast process on N transformation, NIs can reduce NO$_3$ leaching (Lam et al., 2018). There are several NIs, both from organic and synthetic chemical origin. However, the most widely used and reported for better results of reducing NO$_3$ leaching are nitrapyrin, dicyandiamide (DCD), and 3, 4-dimethyl pyrazole phosphate (DMPP) (Subbarao et al., 2006; Haider et al., 2015). Numerous studies in New Zealand and Australia have revealed that NIs are greatly viable at decreasing high level of NO$_3$ leaching from intensively grazed pasture (De Klein and Eckard, 2008). The application of DCD as NI to the animal urine patches has been found to reduce NO$_3$ leaching by 30−79% and increase annual pasture yield by 0−36% (Clough et al.,

2007; Clough and Condron, 2010). However, Lam et al. (2018) reported that the beneficial effects of NIs can be outweighed by indirect $NH_3$ volatilization. Thus, there are several trade-offs and challenges of equal performance of NIs at varying environmental conditions. Therefore, the hazards of $NO_3$ leaching should be managed by cost-effective and environmentally friendly approaches like decreasing the fertilizer application rates and changing the irrigation strategies (Lazicki and Geisseler, 2016).

### 8.5.2 Introducing cover crops

Cover crops are grown to occupy available N in the soil; these can reduce N losses ($NO_3$ leaching, $N_2O$ emission, etc.) in various cropping system that are specific to the region or area. To mitigate $NO_3$ leaching, introducing cover crops according to the areas is one of the crucial strategy. Logsdon et al. (2002) found that after the main crop harvest, leftover N present in the soil can be trapped by cover crops. The repetition of growing cover crops in the cultivated fields has been observed as a proven strategy to reduce $NO_3$ leaching, for example, Logsdon et al. (2002) found a $NO_3$ leaching reduction of 37 kg N $ha^{-1}$ for rye and 60 kg N $ha^{-1}$ for oats. Previous studies suggested that winter cover crops, especially nonlegumes such as grasses and broadleaf species, can reduce $NO_3$ leaching by 35−70% depending on intrinsic (soil and climate) and extrinsic factors (management) (Tonitto et al., 2006; Quemada et al., 2013; Teixeira et al., 2016).

### 8.5.3 Crop rotation

Crop rotation is the practice of growing different kinds of crops in succession in the same area. Understanding the relationship between N and crop rotation is very important when making N management decisions. Crop rotation can play a vital role in reducing the risk of $NO_3$ leaching by efficiently utilizing soil N. Crop rotation also may influence the rate of N mineralization by modifying soil moisture, soil temperature, pH, plant residue, and tillage practices. Studies on the influence of long-term crop rotation by including leguminous crops demonstrated decreasing N fertilizer input and also minimizing the risk of nitrogen leaching, compared with the continuous planting of corn alone. De Notaris et al. (2018) in a 4-year field study found that crop rotation by using legume-based catch crops can minimize N leaching by an average of 23 kg N per hectare in a year and total (60%) across the 4 years. Summer rice−winter wheat rotation is the utmost common pattern of farming in Asia which is receiving the huge synthetic N fertilizer inputs but have lower NUE (Ju et al., 2009; Liang et al., 2011). In wheat−rice rotation, a cycle of dry−wet favors the conditions for leaching of $NO_3$ into shallow groundwater (Liu et al., 2010). When the soil is reflooded before rice seedling transplantation in the following season after the wheat, the $NO_3$ accumulation within the soil is susceptible to leach down. The use of a legume cover in crop rotation can provide a substantial amount of N to a succeeding crop which can help in minimizing N application rate and its losses. Introducing cover crops following the main crop will help to reduce NO$_3$ leaching (Gabriel et al., 2012). Such crops can be placed strategically in a landscape at target locations to reduce NO$_3$ run-off.

### 8.5.4 Application of biochar

Biochar has been proposed as a strategy for reducing N losses and improving agronomic yields on degraded less fertile soils (Laird et al., 2010; Mukherjee et al., 2017). It has been reported in a recent meta-analysis of global biochar studies that biochar can reduce NO$_3$ leaching by 13% with additional benefits of reduced N$_2$O emission (Borchard et al., 2019). It has been observed that biochar particles can retain more NO$_3$ than NH$_4$ in amendment zone of biochar (Haider et al., 2016). Several other studies either under field or laboratory conditions or sorption studies have extensively reported reduced NO$_3$ leaching (Jin et al., 2016), N immobilization (DeLuca et al., 2015), greater N retention (Ding et al., 2010), and sorption to biochar (Yao et al., 2012). In another 4-year field study under temperate sandy soil conditions, Haider et al. (2017) found significant reduction in NO$_3$ leaching due to the application of pinewood pruning biochar in the topsoil (0−15 cm). The yield benefits have been found only in low organic matter containing soils, like in Pakistan.

### 8.5.5 Optimal placement methods and rate of fertilizer application

The large amount and inappropriate fertilizer application methods leads to low NUE around the world (Cui et al., 2007; Hofmeier et al., 2015). It has been observed under field studies that improved fertilization methods like deep placement, root zone application can improve NUE, maintain high stable yield, and contribute toward reducing NO$_3$ leaching (Jiang et al., 2018). For example, deep placement of N fertilizer in a rice crop resulted in 62% more N uptake compared to surface applied N fertilizer. Furthermore, deep placement of N in the root zone can significantly reduce NH$_3$ volatilization (Yao et al., 2018), denitrification (Cai et al., 2002), and leaching (Rochette et al., 2013).

### 8.5.6 Use of slow and controlled release fertilizer

The slow and control release N fertilizers have been developed with a purpose to prolong the N availability for plants uptake. The fertilizer granules are coated with specialized semipermeable or impermeable coatings and water insoluble materials for certain time after application in the field to synchronize the N release and active plant uptake time (Fan and Li, 2010). In this way, the ammonium availability to nitrifiers is reduced and ultimately reduced N losses (Subbarao et al., 2012). Slow and controlled release fertilizers are environment friendly and reduce N losses (Qiao et al., 2016). Application of gypsum-coated urea in different proportions prolonged the fertilizer effect period as well as reduced the N losses (Zhao-Liang, 2000). In another study, Venterea et al. (2011) used two different fertilizers and methods like conventional split application of soluble fertilizers and single application of polymer-coated urea in a field study and found substantial reduction in $NO_3$ leaching when polymer-coated urea was used.

## 8.6 Options for N₂O mitigation

Nitrous oxide is one of the most powerful greenhouse gases (GHGs) in the atmosphere and $\sim 40\%$ of $N_2O$ comes from anthropogenic activities like forest burning, agricultural practices, N fertilizer application, and several industrial processes (Zellweger et al., 2019). Biological processes such as nitrification and denitrification are responsible for $N_2O$ emission. Some of the major management practices to reduce $N_2O$ emission are discussed below.

### 8.6.1 Use of enhanced efficiency fertilizers

Nitrogen fertilizer application is linked with the emission of $N_2O$, as N content is directly proportional to nitrification and denitrification (Akiyama et al., 2000; Passianoto et al., 2003; Zanatta et al., 2010; Signor and Cerri, 2013). The nitrification process will increase with the addition of the higher amount of fertilizer having $N-NH_4^+$ (Mosier, 2001; Khalil et al., 2004). With the high application of $NO_3$, $N_2O$ emission will be higher and vice versa, as $NO_3$ is responsible for denitrification (do Carmo et al., 2005; Ruser et al., 2006; Hellebrand et al., 2008). High application of $NH_4NO_3$ in the field of sugarcane induced more $N_2O$ emission and $N_2O$ production by urea reached a maximum of $114$ kg N ha$^{-1}$ (Signor et al., 2013). Application of $N-NO_3$ and $N-NH_4$ fertilizers to the soil enhanced production of $N_2O$ by $56\%$ and $15\%$, respectively (DeLaune et al., 1998). Reduction in the application rates of N fertilizers is widely recognized as the most effective measure of reducing $N_2O$ emissions. In addition, use of enhanced efficiency fertilizers having the property of slow releasing N to the soil could be quite effective to control $N_2O$ emission (Shaviv, 2001). In Brazil, slow-release N fertilizer maintained lower production of $N_2O$ as compared to calcium nitrate, urea, ammonium nitrate, and ammonium sulfate fertilizers (Zanatta et al., 2010). Similarly, use of NIs has great potential in mitigating $N_2O$ emissions with a mean reduction of $38\%$ compared with conventional fertilizers (Akiyama et al., 2010).

### 8.6.2 Soil tillage

Soil tillage practices directly affect soil properties such as soil eration, residue decomposition rate, soil temperature, soil structure, microbial activity, N mineralization, and soil moisture, responsible for enhanced $N_2O$ production. In literature, there are mixed responses of $N_2O$ emissions to reduced tillage (RT)/no-till (NT) depending on soil types and regions and crop residue management (Mutegi et al., 2010). For instance, under NT, $N_2O$ production was increased threefolds compared to conventional tillage (CT) management system in Brazil (Escobar et al., 2010).

Mutegi et al. (2010) demonstrated that there was no impact of different tillage practices when residue was not incorporated in soil but in residue retention scenario, CT resulted in higher $N_2O$ emissions compared with NT. So, NT or RT has the potential for reducing $N_2O$ emissions when crop residues are returned to the soil, especially in light textured soils.

### 8.6.3 Soil amendment with biochar

The field-scale research information on GHGs emission from agricultural crops is scarce in Pakistan. Majority of available data are secondary by making calculations from per capita energy consumption or other resources, for example, Mir et al. (2017) and Iqbal and Goheer (2008). There are few studies where real-time field-scale GHGs, particularly $N_2O$ flux measurements have been performed and reported. For instance, Mahmood et al. (2008) reported a year-long field study to monitor $N_2O$ emission from irrigated cotton where they found low $N_2O$ emission even with high soil moisture and temperature. Recently, Dawar et al. (2021) used $^{15}N$ tracing techniques by applying NI, mulches, and biochar to reduce field-scale $N_2O$ emission from cultivated wheat in KPK Pakistan. They found that application of nitropyrene, biochar, and mulches have potential to reduce ammonia and $N_2O$ emission. Many studies have shown that biochar addition in soil could be an effective approach to improve water and nutrient retention, reduce $N_2O$ emission, and promote crop production. According to a meta-analysis study, by biochar application, there were about 54% reduction of $N_2O$ emission (Cayuela et al., 2014). But biochar effectiveness for mitigation of $N_2O$ emission depends on its production method, pyrolysis temperature, and feedstock type because these factors determine biochar biochemical properties (Butnan et al., 2016).

## 8.7 On-farm technologies and practices to improve NUE

The best among worldwide traditionally used practices to manage NUE are cultivation of high yielding crop varieties, calculated and recommended fertilizer application, early sowing of winter crops avoiding seedbed application of nitrogen fertilizer, split application of fertilizers in spring, crop rotation, permitting mineralized N from the residue of the previous crop, organic farming, using leguminous crop residues and animal manure. Advance technologies include use of remotely sensed images through satellite and internet-accessibly N management tools (Kitchen et al., 2001).

Development of technologies to enhance NUE of applied fertilizers is the need of time. Along with the various traditional approaches, currently the researchers are focusing on the coating or impregnating plant growth promoting rhizobacteria on chemical fertilizers to increase NUE. Biofertilizer technology not only improves the NUE but also minimizes the additional cost and efforts of end-user putting for the application of biofertilizers along with the agrochemicals.

### 8.7.1 Optimum N fertilizer management

In Pakistan, majority of farmers have small land holding and primarily rely on high N and moderate to low P application for average to below average crop production. Khan et al. (2010) reported that crop productivity decreased with the increase in prices of agricultural inputs. Optimum crop growth and development requires balanced application of nutrients. Nitrogen is present 2–4% in healthy plants, less than this proportion results in deficiency of plant protein formation and ultimately causes chlorosis (Brady et al., 2008) and stunted plant growth (Bingham et al., 2012).

### 8.7.2 Animal and poultry manures application

Along with the other traditional strategies, application of animal and poultry manures is also an effective way to enhance soil fertility and to improve NUE. These organic manures are great source of nutrients and can enhance N availability by enhancing soil exchange capacity. Poultry manure can also be used as nutrients source, especially organic N for crop production. If poultry manure is properly managed in Pakistan, it will contribute 101,000 tons of N, 58,000 tons of $P_2O_5$, and 26,000 tons of $K_2O$ (FAO, 2004).

### 8.7.3 Green manuring

Green manure is usually considered as legume-based cover crop sown to enhance the fertility of the soil between two crops demanding high nitrogen inputs (Thomas et al., 2019). Use of green manure crops makes it possible to provide 40–60% of the total N requirement of the successive crops. After incorporation of these leguminous crops in soil, organic N is decomposed gradually and mineral N is available to plants for longer growth period (Fig. 8.4). Yang et al. (2018) tested the effect of green manure crop cultivation on soil N fractions and nutrient uptake in their 4-year field study using maize as successive crop. Green manure incorporation resulted in increased maize yield with reduced input of synthetic fertilizer. According to Prakash and Bhushan (2003), 3% N present in $3.14 \, t \, ha^{-1}$ Leucaena leaves increased wheat yield as compared to N applied in urea. Similarly, Leucaena leaves having 3.83–4.25% N are used as N supply ($60 \, kg \, N \, ha^{-1}$) in maize–wheat cropping system and showed increase in wheat yield as compared to equal quantity of urea (Sharma and Behera, 2010).

### 8.7.4 Utilization of genetic tools for NUE

Besides agronomic approaches, genetic manipulations can be adapted as a promising tool for the development of crop varieties which are highly efficient in N uptake, metabolism, and assimilation (Nguyen and Kant, 2018). Recent genomic studies have revealed the significant role of various root to shoot signaling pathways, synthesis of amino acids, and feedback mechanism regulation for nitrogen uptake

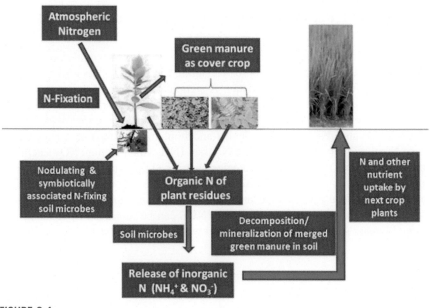

**FIGURE 8.4**

Role of green manure crops in management of soil nitrogen.

and assimilation (Tiwari et al., 2017). Overexpression of GS1 gene in wheat improved the dry biomass and grains yield under control of multiunit promoter sequence including RolD, CAM V 35S, and rbcS through expansion of root system and photosynthetic activity (Good et al., 2007; Moreau et al., 2012). Literature supports the upregulation of GS GOGAT genes to improve crop yield by modulating NUE (Tremblay et al., 2012).

### 8.7.5 Introduction of mix farming

Mixed farming combines livestock and cropping at farm and landscape scales. It provides opportunities to connect N inputs and surpluses, with the aim of reducing overall levels of N pollution and of increasing NUE. The opposite can be illustrated by the situation where arable farming areas export grain to livestock farming areas, leading to excess manure in the livestock areas that cannot be used locally. Combining cropping and livestock locally can therefore help reduce pollution (Wilkins, 2008; Sutton et al., 2013). The goal is to achieve an optimized distribution of manure and fodder import/production between fields and farms (Garrett et al., 2017; Asai et al., 2018). The reconnection of crop and livestock increases the overall landscape-level NUE and has been demonstrated to reduce N surplus and water pollution (Garnier et al., 2016).

## 8.8 Conclusions: toward managing agricultural soils to mitigate N losses

Nitrogen is the most essential nutrient for plant growth but its excessive application and low NUE are the main reasons for economic losses as well as environmental pollution worldwide, including Pakistan. A significant portion of the applied N is not utilized by plants due to N losses such as leaching, denitrification, erosion, fixation, runoff, or volatilization. It requires enhanced NUE in agriculture for economic and environmental reasons and to devise and implement mitigation strategies to reduce N pollution. In order to mitigate N losses ($NH_3$ volatilization, $NO_3$ leaching, and $N_2O$ emissions), systematic strategies such as (1) appropriate N source, (2) application methods and time, (3) use of urease and NIs, (4) use of cover crops in crop rotations, (5) use of biochar and other organic amendments, and (6) green manuring should be adopted. Furthermore, NUE can be improved by reducing or reusing biomass streams in the form of food waste, manure, and sewage.

## References

Abbasi, M.K., Tahir, M.M., Rahim, N., 2013. Effect of N fertilizer source and timing on yield and N use efficiency of rainfed maize (*Zea mays* L.) in Kashmir—Pakistan. Geoderma 195, 87—93.

Akiyama, H., Tsuruta, H., Watanabe, T., 2000. $N_2O$ and NO emissions from soils after the application of different chemical fertilizers. Chemosphere Global Change Sci. 2, 313—320.

Akiyama, H., Yan, X., Yagi, K., 2010. Evaluation of effectiveness of enhanced-efficiency fertilizers as mitigation options for $N_2O$ and NO emissions from agricultural soils: meta-analysis. Global Change Biol. 16, 1837—1846.

Alia, S., Irshadb, M., Khanc, N., Muhammadd, Z., 2016. Effects of fertilizers on incidence of damping-off in nursery of flue cured Virginia cv. speight g-28. J. Biosc. Agric. Res. 10, 877—885.

Asai, M., Moraine, M., Ryschawy, J., de Wit, J., Hoshide, A.K., Martin, G., 2018. Critical factors for crop-livestock integration beyond the farm level: a cross-analysis of worldwide case studies. Land Use Pol. 73, 184—194.

Bellarby, J., Tirado, R., Leip, A., Weiss, F., Lesschen, J.P., Smith, P., 2013. Livestock greenhouse gas emissions and mitigation potential in Europe. Global Change Biol. 19, 3—18.

Biernat, L., Taube, F., Vogeler, I., Reinsch, T., Kluß, C., Loges, R., 2020. Is organic agriculture in line with the EU-Nitrate directive? On-farm nitrate leaching from organic and conventional arable crop rotations. Agric. Ecosyst. Environ. 298, 106964.

Bingham, I., Karley, A., White, P., Thomas, W., Russell, J., 2012. Analysis of improvements in nitrogen use efficiency associated with 75 years of spring barley breeding. Eur. J. Agron. 42, 49—58.

Bittman, S., Dedina, M., Howard, C., Oenema, O., Sutton, M., 2014. Options for Ammonia Mitigation: Guidance From the Unece Task Force on Reactive Nitrogen. NERC/Centre for Ecology and Hydrology.

Borchard, N., Schirrmann, M., Cayuela, M.L., Kammann, C., Wrage-Mönnig, N., Estavillo, J.M., Fuertes-Mendizábal, T., Sigua, G., Spokas, K., Ippolito, J.A., 2019. Biochar, soil and land-use interactions that reduce nitrate leaching and $N_2O$ emissions: a meta-analysis. Sci. Total Environ. 651, 2354−2364.

Bouwman, L., Daniel, J.S., Davidson, E.A., de Klein, C., Holland, E., Ju, X., Kanter, D., Oenema, O., Ravishankara, A., Skiba, U.M., 2013. Drawing Down $N_2O$ to Protect Climate and the Ozone Layer. A UNEP Synthesis Report. United Nations Environment Programme (UNEP).

Brady, N.C., Weil, R.R., Weil, R.R., 2008. The Nature and Properties of Soils. Prentice Hall, Upper Saddle River, NJ.

Broderick, G., 2003. Effects of varying dietary protein and energy levels on the production of lactating dairy cows. J. Dairy Sci. 86, 1370−1381.

Bruulsema, T., Lemunyon, J., Herz, B., 2009. Know your fertilizer rights. Crop Soils Mag. 42, 13−18.

Burton, C.H., 2007. The potential contribution of separation technologies to the management of livestock manure. Livest. Sci. 112, 208−216.

Butnan, S., Deenik, J.L., Toomsan, B., Antal, M.J., Vityakon, P., 2016. Biochar properties influencing greenhouse gas emissions in tropical soils differing in texture and mineralogy. J. Environ. Qual. 45, 1509−1519.

Butterbach-Bahl, K., Nemitz, E., Zaehle, S., 2011. Effect of reactive nitrogen on the European greenhouse balance. In: Sutton, M.A., Howard, C., Erisman, J.W., Billen, G., Bleeker, A., Grenfelt, P., van Grinsven, H., Grizzetti, B. (Eds.), The European Nitrogen Assessment. Cambridge University Press, pp. 434−462. Chapter 19.

Cai, G., Chen, D., Ding, H., Pacholski, A., Fan, X., Zhu, Z., 2002. Nitrogen losses from fertilizers applied to maize, wheat and rice in the North China Plain. Nutr. Cycl. Agroecosyst. 63, 187−195.

Cayuela, M., Van Zwieten, L., Singh, B., Jeffery, S., Roig, A., Sánchez-Monedero, M., 2014. Biochar's role in mitigating soil nitrous oxide emissions: a review and meta-analysis. Agric. Ecosyst. Environ. 191, 5−16.

Clough, T.J., Condron, L.M., 2010. Biochar and the nitrogen cycle: introduction. J. Environ. Qual. 39, 1218−1223.

Clough, T.J., Di, H.J., Cameron, K.C., Sherlock, R.R., Metherell, A., Clark, H., Rys, G., 2007. Accounting for the utilization of a $N_2O$ mitigation tool in the IPCC inventory methodology for agricultural soils. Nutr. Cycl. Agroecosyst. 78, 1−14.

Čuček, L., Klemeš, J.J., Kravanja, Z., 2012. A review of footprint analysis tools for monitoring impacts on sustainability. J. Clean. Prod. 34, 9−20.

Cui, Z.-L., Chen, X.-P., Zhang, F.-S., Xu, J.-F., Shi, L.-W., Li, J.-L., 2007. Appropriate soil nitrate N content for a winter wheat/summer maize rotation system in North China Plain. J. Appl. Ecol. 18, 2227−2232.

Davies, D., 2000. The nitrate issue in England and Wales. Soil Use Manag. 16, 142−144.

Dawar, K., Sardar, K., Zaman, M., Mueller, C., Alberto, S.-C., Aamir, K., Borzouei, A., Pérez-Castillo, A.G., 2021. Effects of the nitrification inhibitor nitrapyrin and the plant growth regulator gibberellic acid on yield-scale nitrous oxide emission in maize fields under hot climatic conditions. Pedosphere 31, 323−331.

De Klein, C., Eckard, R., 2008. Targeted technologies for nitrous oxide abatement from animal agriculture. Aust. J. Exp. Agric. 48, 14−20.

De Notaris, C., Rasmussen, J., Sørensen, P., Olesen, J.E., 2018. Nitrogen leaching: a crop rotation perspective on the effect of N surplus, field management and use of catch crops. Agric. Ecosyst. Environ. 255, 1−11.

DeLaune, R., Pezeshki, S., Lindau, C., 1998. Influence of soil redox potential on nitrogen uptake and growth of wetland oak seedlings. J. Plant Nutr. 21, 757−768.

DeLuca, T.H., Gundale, M.J., MacKenzie, M.D., Jones, D.L., 2015. Biochar effects on soil nutrient transformations. In: Biochar for Environmental Management: Science, Technology and Implementation, vol. 2, pp. 421−454.

Di, H., Cameron, K., Sherlock, R., 2007. Comparison of the effectiveness of a nitrification inhibitor, dicyandiamide, in reducing nitrous oxide emissions in four different soils under different climatic and management conditions. Soil Use Manag. 23, 1−9.

Ding, Y., Liu, Y.-X., Wu, W.-X., Shi, D.-Z., Yang, M., Zhong, Z.-K., 2010. Evaluation of biochar effects on nitrogen retention and leaching in multi-layered soil columns. Water Air Soil Pollut. 213, 47−55.

do Carmo, J.B., Neill, C., Garcia-Montiel, D.C., de Cássia Piccolo, M., Cerri, C.C., Steudler, P.A., de Andrade, C.A., Passianoto, C.C., Feigl, B.J., Melillo, J.M., 2005. Nitrogen dynamics during till and no-till pasture restoration sequences in Rondônia, Brazil. Nutr. Cycl. Agroecosyst. 71, 213−225.

Elliott, H., Collins, N., 1982. Factors affecting ammonia release in broiler houses. Transactions ASAE 25, 413−0418.

Engel, R., Jones, C., Wallander, R., 2011. Ammonia volatilization from urea and mitigation by NBPT following surface application to cold soils. Soil Sci. Soc. Am. J. 75, 2348−2357.

Erisman, J.W., Sutton, M.A., Galloway, J., Klimont, Z., Winiwarter, W., 2008. How a century of ammonia synthesis changed the world. Nat. Geosci. 1, 636−639.

Escobar, L.F., Amado, T.J.C., Bayer, C., Chavez, L.F., Zanatta, J.A., Fiorin, J.E., 2010. Post-harvest nitrous oxide emissions from a subtropical oxisol as influenced by summer crop residues and their management. Rev. Bras. Cienc. Solo. 34, 435−442.

Fan, X., Li, Y., 2010. Nitrogen release from slow-release fertilizers as affected by soil type and temperature. Soil Sci. Soc. Am. J. 74, 1635−1641.

Fangueiro, D., Coutinho, J., Chadwick, D., Moreira, N., Trindade, H., 2008a. Effect of cattle slurry separation on greenhouse gas and ammonia emissions during storage. J. Environ. Qual. 37, 2322−2331.

Fangueiro, D., Pereira, J., Chadwick, D., Coutinho, J., Moreira, N., Trindade, H., 2008b. Laboratory assessment of the effect of cattle slurry pre-treatment on organic N degradation after soil application and $N_2O$ and $N_2$ emissions. Nutr. Cycl. Agroecosyst. 80, 107−120.

Fangueiro, D., Hjorth, M., Gioelli, F., 2015. Acidification of animal slurry−a review. J. Environ. Manag. 149, 46−56.

FAO, 2004. Fertilizer Use by Crop in Pakistan. http://www.fao.org/3/y5460e/y5460e00.htm#Contents.

Fenn, L., Hossner, L., 1985. Ammonia volatilization from ammonium or ammonium-forming nitrogen fertilizers. Adv. Soil Sci. 123−169. Springer.

Foster, G.D., Roberts Jr., E.C., Gruessner, B., Velinsky, D.J., 2000. Hydrogeochemistry and transport of organic contaminants in an urban watershed of Chesapeake Bay (USA). Appl. Geochem. 15, 901−915.

Fowler, D., Coyle, M., Skiba, U., Sutton, M.A., Cape, J.N., Reis, S., Sheppard, L.J., Jenkins, A., Grizzetti, B., Galloway, J.N., 2013a. The global nitrogen cycle in the twenty-first century. Philos. T. Roy. Soc. B 368, 20130164.

Fowler, D., Coyle, M., Skiba, U., Sutton, M.A., Cape, J.N., Reis, S., Sheppard, L.J., Jenkins, A., Grizzetti, B., Galloway, J.N., Vitousek, P., Leach, A., Bouwman, A.F., Butterbach-Bahl, K., Dentener, F., Stevenson, D., Amann, M., Voss, M., 2013b. The global nitrogen cycle in the twenty-first century. Phil. Trans. Biol. Sci. 368, 20130164.

Furtado da Silva, N., Cabral da Silva, E., Muraoka, T., Batista Teixeira, M., Antonio Loureiro Soares, F., Nobre Cunha, F., Adu-Gyamfi, J., Cavalcante, W.S.d.S., 2020. Nitrogen utilization from ammonium nitrate and urea fertilizer by irrigated sugarcane in Brazilian Cerrado Oxisol. Agriculture 10, 323.

Gabriel, J., Muñoz-Carpena, R., Quemada, M., 2012. The role of cover crops in irrigated systems: water balance, nitrate leaching and soil mineral nitrogen accumulation. Agric. Ecosyst. Environ. 155, 50—61.

Galloway, J.N., Aber, J.D., Erisman, J.W., Seitzinger, S.P., Howarth, R.W., Cowling, E.B., Cosby, B.J., 2003. The nitrogen Cascade. BioScience 53, 341—356.

Galloway, J.N., Townsend, A.R., Erisman, J.W., Bekunda, M., Cai, Z., Freney, J.R., Martinelli, L.A., Seitzinger, S.P., Sutton, M.A., 2008. Transformation of the nitrogen cycle: recent trends, questions, and potential solutions. Science 320, 889—892.

Galloway, J.N., Winiwarter, W., Leip, A., Leach, A.M., Bleeker, A., Erisman, J.W., 2014. Nitrogen footprints: past, present and future. Environ. Res. Lett. 9, 115003.

Garnier, J., Anglade, J., Benoit, M., Billen, G., Puech, T., Ramarson, A., Passy, P., Silvestre, M., Lassaletta, L., Trommenschlager, J.-M., 2016. Reconnecting crop and cattle farming to reduce nitrogen losses to river water of an intensive agricultural catchment (Seine basin, France): past, present and future. Environ. Sci. Pol. 63, 76—90.

Garrett, R., Niles, M.T., Gil, J.D., Gaudin, A., Chaplin-Kramer, R., Assmann, A., Assmann, T.S., Brewer, K., de Faccio Carvalho, P.C., Cortner, O., 2017. Social and ecological analysis of commercial integrated crop livestock systems: current knowledge and remaining uncertainty. Agric. Syst. 155, 136—146.

Gavrilova, O., Leip, A., Dong, H., MacDonald, J.D., Gomez Bravo, C., Amon, B., Barahona Rosales, R., Prado, A.d., de Lima, M.A., Oyhantçabal, W., 2019. Emmisions from Livestock and Manure Management. Embrapa Meio Ambiente-Capítulo em livro científico (ALICE).

Gilhespy, S.L., Webb, J., Chadwick, D.R., Misselbrook, T.H., Kay, R., Camp, V., Retter, A.L., Bason, A., 2009. Will additional straw bedding in buildings housing cattle and pigs reduce ammonia emissions? Biosyst. Eng. 102, 180—189.

Godfray, H.C.J., Beddington, J.R., Crute, I.R., Haddad, L., Lawrence, D., Muir, J.F., Pretty, J., Robinson, S., Thomas, S.M., Toulmin, C., 2010. Food security: the challenge of feeding 9 billion people. Science 327, 812—818.

Good, A.G., Johnson, S.J., De Pauw, M., Carroll, R.T., Savidov, N., Vidmar, J., Lu, Z., Taylor, G., Stroeher, V., 2007. Engineering nitrogen use efficiency with alanine aminotransferase. Botany 85, 252—262.

Gu, B., Leach, A.M., Ma, L., Galloway, J.N., Chang, S.X., Ge, Y., Chang, J., 2013. Nitrogen footprint in China: food, energy, and nonfood goods. Environ. Sci. Technol. 47, 9217—9224.

Haider, G., Koyro, H.-W., Azam, F., Steffens, D., Müller, C., Kammann, C., 2015. Biochar but not humic acid product amendment affected maize yields via improving plant-soil moisture relations. Plant Soil 395, 141—157.

Haider, G., Steffens, D., Müller, C., Kammann, C.I., 2016. Standard extraction methods may underestimate nitrate stocks captured by field-aged biochar. J. Environ. Qual. 45, 1196—1204.

Haider, G., Steffens, D., Moser, G., Müller, C., Kammann, C.I., 2017. Biochar reduced nitrate leaching and improved soil moisture content without yield improvements in a four-year field study. Agric. Ecosyst. Environ. 237, 80–94.

Haroon, M., Idrees, F., Naushahi, H.A., Afzal, R., Usman, M., Qadir, T., Rauf, H., 2019. Nitrogen use efficiency: farming practices and sustainability. J. Exp. Agric. Int. 1–11.

Hatfield, J.L., Follett, R.F., 2008. Nitrogen in the Environment: Sources, Problems and Management. Elsevier, Academic Press.

Heffer, P., Prud'homme, M., 2016. Global nitrogen fertilizer demand and supply: trend, current level and outlook. In: International Nitrogen Initiative Conference. Melbourne, Australia.

Hellebrand, H., Scholz, V., Kern, J., 2008. Nitrogen conversion and nitrous oxide hot spots in energy crop cultivation. Res. Agric. Engineer. 54, 58–67.

Hofmeier, M., Roelcke, M., Han, Y., Lan, T., Bergmann, H., Böhm, D., Cai, Z., Nieder, R., 2015. Nitrogen management in a rice–wheat system in the Taihu Region: recommendations based on field experiments and surveys. Agric. Ecosyst. Environ. 209, 60–73.

Iqbal, M.M., Goheer, M.A., 2008. Greenhouse gas emissions from agro-ecosystems and their contribution to environmental change in the Indus Basin of Pakistan. Adv. Atmos. Sci. 25, 1043–1052.

Jahangir, M.M., Khalil, M.I., Johnston, P., Cardenas, L., Hatch, D., Butler, M., Barrett, M., O'flaherty, V., Richards, K.G., 2012. Denitrification potential in subsoils: a mechanism to reduce nitrate leaching to groundwater. Agric. Ecosyst. Environ. 147, 13–23.

Jensen, L.S., 2013. Animal manure residue upgrading and nutrient recovery in biofertilisers. In: Animal Manure Recycling: Treatment and Management, pp. 271–294.

Jiang, C., Lu, D., Zu, C., Zhou, J., Wang, H., 2018. Root-zone fertilization improves crop yields and minimizes nitrogen loss in summer maize in China. Sci. Rep. 8, 1–9.

Jin, Z., Chen, X., Chen, C., Tao, P., Han, Z., Zhang, X., 2016. Biochar impact on nitrate leaching in upland red soil, China. Environ. Earth Sci. 75, 1–10.

Jones, C., Brown, B.D., Engel, R., Horneck, D., Olson-Rutz, K., 2013. Nitrogen Fertilizer Volatilization. Montana State University Extension, p. EBO208.

Ju, X.-T., Xing, G.-X., Chen, X.-P., Zhang, S.-L., Zhang, L.-J., Liu, X.-J., Cui, Z.-L., Yin, B., Christie, P., Zhu, Z.-L., 2009. Reducing environmental risk by improving N management in intensive Chinese agricultural systems. Proc. Natl. Acad. Sci. U. S. A 106, 3041–3046.

Khalil, K., Mary, B., Renault, P., 2004. Nitrous oxide production by nitrification and denitrification in soil aggregates as affected by O2 concentration. Soil Biol. Biochem. 36, 687–699.

Khan, H.G.A., Ahmad, A., e Siraj, A., 2010. Impact of rising prices of fertilizers on crops production in Pakistan. Global J. Manag. Bus. 10.

Kitchen, N., Goulding, K., Follett, R., Hatfield, J., 2001. On-farm technologies and practices to improve nitrogen use efficiency. In: Nitrogen in the Environment: Sources, Problems and Management. Elsevier, Amsterdam, The Netherlands.

Kocatürk-Schumacher, N.P., Bruun, S., Zwart, K., Jensen, L.S., 2017. Nutrient recovery from the liquid fraction of digestate by clinoptilolite. Clean 45, 1500153.

Kocatürk-Schumacher, N.P., Zwart, K., Bruun, S., Stoumann Jensen, L., Sørensen, H., Brussaard, L., 2019. Recovery of nutrients from the liquid fraction of digestate: use of enriched zeolite and biochar as nitrogen fertilizers. J. Plant Nutr. Soil Sci. 182, 187–195.

Laird, D., Fleming, P., Wang, B., Horton, R., Karlen, D., 2010. Biochar impact on nutrient leaching from a Midwestern agricultural soil. Geoderma 158, 436–442.

Lam, S.K., Suter, H., Davies, R., Bai, M., Mosier, A.R., Sun, J., Chen, D., 2018. Direct and indirect greenhouse gas emissions from two intensive vegetable farms applied with a nitrification inhibitor. Soil Biol. Biochem. 116, 48—51.

Lawson, G., Bealey, W.J., Dupraz, C., Skiba, U.M., 2020. Agroforestry and opportunities for improved nitrogen management. In: Just Enough Nitrogen. Springer, pp. 393—417.

Lazicki, P., Geisseler, D., 2016. Soil nitrate testing supports nitrogen management in irrigated annual crops. Calif. Agric. 71, 90—95.

Leach, A.M., Galloway, J.N., Bleeker, A., Erisman, J.W., Kohn, R., Kitzes, J., 2012. A nitrogen footprint model to help consumers understand their role in nitrogen losses to the environment. Environ. Dev. 1, 40—66.

Liang, Y., Li, Y., Wang, H., Zhou, J., Wang, J., Regier, T., Dai, H., 2011. Co 3 O 4 nanocrystals on graphene as a synergistic catalyst for oxygen reduction reaction. Nat. Mater. 10, 780—786.

Liu, C., Zheng, X., Zhou, Z., Han, S., Wang, Y., Wang, K., Liang, W., Li, M., Chen, D., Yang, Z., 2010. Nitrous oxide and nitric oxide emissions from an irrigated cotton field in Northern China. Plant Soil 332, 123—134.

Liu, J., Ma, K., Ciais, P., Polasky, S., 2016. Reducing human nitrogen use for food production. Sci. Rep. 6, 30104.

Liu, Q., Liu, B., Zhang, Y., Hu, T., Lin, Z., Liu, G., Wang, X., Ma, J., Wang, H., Jin, H., Ambus, P., Amonette, J.E., Xie, Z., 2019. Biochar application as a tool to decrease soil nitrogen losses ($NH_3$ volatilization, $N_2O$ emissions, and N leaching) from croplands: options and mitigation strength in a global perspective. Global Change Biol. 25, 2077—2093.

Logsdon, S., Kaspar, T.C., Meek, D.W., Prueger, J.H., 2002. Nitrate leaching as influenced by cover crops in large soil monoliths. Agron. J. 94, 807—814.

Lukin, S.M., Nikolskiy, K.S., Ryabkov, V.V., Rysakova, I.V., 2014. Methods to reduce ammonia nitrogen losses during production and application of organic fertilizers. In: Ammonia Workshop 2012. Abating Ammonia Emissions in the UNECE and EECCA region, Saint Petersburg, pp. 169—175.

Mahmood, T., Ali, R., Iqbal, J., Robab, U., 2008. Nitrous oxide emission from an irrigated cotton field under semiarid subtropical conditions. Biol. Fertil. Soils 44, 773—781.

McKervey, Z., Woods, V., Easson, D., Forbes, E., 2005. Opportunities to Reduce Nitrate Leaching From Grazed Grassland A Summary of Research Findings in New Zealand. Global Research Unit, AFBI, Hillborough, Northern Ireland.

Melse, R.W., Ogink, N., 2005. Air scrubbing techniques for ammonia and odor reduction at livestock operations: review of on-farm research in The Netherlands. Transactions ASAE 48, 2303—2313.

Melse, R., Ogink, N., Bosma, A., 2008. Multi-pollutant scrubbers for removal of ammonia, odor, and particulate matter from animal house exhaust air. In: Exploring the Advantages, Limitations, and Economics of Mitigation Technologies.

Melse, R., Hofschreuder, P., Ogink, N., 2012. Removal of particulate matter (PM10) by air scrubbers at livestock facilities: results of an on-farm monitoring program. Transactions ASABE 55, 689—698.

Mir, K.A., Purohit, P., Mehmood, S., 2017. Sectoral assessment of greenhouse gas emissions in Pakistan. Environ. Sci. Pollut. Res. 24, 27345—27355.

Misselbrook, T.H., Powell, J.M., 2005. Influence of bedding material on ammonia emissions from cattle excreta. J. Dairy Sci. 88, 4304—4312.

Modolo, L.V., da-Silva, C.J., Brandão, D.S., Chaves, I.S., 2018. A minireview on what we have learned about urease inhibitors of agricultural interest since mid-2000s. J. Adv. Res. 13, 29–37.

Moller, B., Hansen, D., Sorensen, C., 2007. Nutrient recovery by solid-liquid separation and methane productivity of solids. Transactions ASABE 50, 193–200.

Moreau, M., Azzopardi, M., Clément, G., Dobrenel, T., Marchive, C., Renne, C., Martin-Magniette, M.-L., Taconnat, L., Renou, J.-P., Robaglia, C., 2012. Mutations in the Arabidopsis homolog of LST8/GβL, a partner of the target of rapamycin kinase, impair plant growth, flowering, and metabolic adaptation to long days. Plant Cell 24, 463–481.

Mosier, A.R., 2001. Exchange of gaseous nitrogen compounds between agricultural systems and the atmosphere. Plant Soil 228, 17–27.

Mukherjee, R., Basu, J., Mandal, P., Guha, P.K., 2017. A review of micromachined thermal accelerometers. J. Micromech. Microeng. 27, 123002.

Mutegi, J.K., Munkholm, L.J., Petersen, B.M., Hansen, E.M., Petersen, S.O., 2010. Nitrous oxide emissions and controls as influenced by tillage and crop residue management strategy. Soil Biol. Biochem. 42, 1701–1711.

Nguyen, G.N., Kant, S., 2018. Improving nitrogen use efficiency in plants: effective phenotyping in conjunction with agronomic and genetic approaches. Funct. Plant Biol. 45, 606–619.

Oenema, O., Bleeker, A., Braathen, N.A., Budňakova, M., Bull, K., Čermak, P., Geupel, M., Hicks, K., Hoft, R., Kozlova, N., 2011. Nitrogen in Current European Policies. The European nitrogen assessment., pp. 62–81

Ogink, N.W., Bosma, B.J., 2007. Multi-phase air scrubbers for the combined abatement of ammonia, odor and particulate matter emissions. In: International Symposium on Air Quality and Waste Management for Agriculture, 16–19 September 2007. American Society of Agricultural and Biological Engineers, Broomfield, Colorado, p. 37.

Oita, A., Malik, A., Kanemoto, K., Geschke, A., Nishijima, S., Lenzen, M., 2016. Substantial nitrogen pollution embedded in international trade. Nat. Geosci. 9, 111–115.

Oita, A., Nagano, I., Matsuda, H., 2018. Food nitrogen footprint reductions related to a balanced Japanese diet. AMBIO 47, 318–326.

Pan, B., Lam, S.K., Mosier, A., Luo, Y., Chen, D., 2016. Ammonia volatilization from synthetic fertilizers and its mitigation strategies: a global synthesis. Agric. Ecosyst. Environ. 232, 283–289.

Passianoto, C.C., Ahrens, T., Feigl, B.J., Steudler, P.A., Do Carmo, J.B., Melillo, J.M., 2003. Emissions of $CO_2$, $N_2O$, and NO in conventional and no-till management practices in Rondônia, Brazil. Biol. Fertil. Soils 38, 200–208.

Patterson, P., 2005. Management strategies to reduce air emissions: emphasis—Dust and ammonia. J. Appl. Poultry Res. 14, 638–650.

Pavlidis, G., Tsihrintzis, V.A., 2018. Environmental benefits and control of pollution to surface water and groundwater by agroforestry systems: a review. Water Resour. Manag. 32, 1–29.

Petersen, S.O., Andersen, A.J., Eriksen, J., 2012. Effects of cattle slurry acidification on ammonia and methane evolution during storage. J. Environ. Qual. 41, 88–94.

Powell, J.M., Misselbrook, T.H., Casler, M.D., 2008. Season and bedding impacts on ammonia emissions from tie-stall dairy barns. J. Environ. Qual. 37, 7–15.

Prakash, O., Bhushan, L., 2003. Effect of fertilizer substitution through white leadtree (*Leucaena leucocephala*) green biomass on growth, yield and economics of wheat (*Triticum aestivum*) crop in degraded lands. Indian J. Agric. Sci. 73, 311–314.

Qiao, D., Liu, H., Yu, L., Bao, X., Simon, G.P., Petinakis, E., Chen, L., 2016. Preparation and characterization of slow-release fertilizer encapsulated by starch-based superabsorbent polymer. Carbohydr. Polym. 147, 146–154.

Qin, S., Hu, C., Zhang, Y., Wang, Y., Dong, W., Li, X., 2011. Advances in nitrogen footprint research. Zhongguo Shengtai Nongye Xuebao 19, 462–467. Chinese Journal of Eco-Agriculture.

Quemada, M., Baranski, M., Nobel-de Lange, M., Vallejo, A., Cooper, J., 2013. Meta-analysis of strategies to control nitrate leaching in irrigated agricultural systems and their effects on crop yield. Agric. Ecosyst. Environ. 174, 1–10.

Raza, S., Zhou, J., Aziz, T., Afzal, M.R., Ahmed, M., Javaid, S., Chen, Z., 2018. Piling up reactive nitrogen and declining nitrogen use efficiency in Pakistan: a challenge not challenged (1961–2013). Environ. Res. Lett. 13, 034012.

Reis, S., Howard, C., Sutton, M.A., 2015. Costs of Ammonia Abatement and the Climate Co-benefits. Springer.

Ritz, C., Mitchell, B., Fairchild, B., Czarick III, M., Worley, J., 2006. Improving in-house air quality in broiler production facilities using an electrostatic space charge system. J. Appl. Poultry Res. 15, 333–340.

Rochette, P., Angers, D.A., Chantigny, M.H., Gasser, M.O., MacDonald, J.D., Pelster, D.E., Bertrand, N., 2013. Ammonia volatilization and nitrogen retention: how deep to incorporate urea? J. Environ. Qual. 42, 1635–1642.

Rogers, C.W., Dari, B., Walsh, O.S., 2018. Best management practices for minimizing ammonia volatilization from fertilizer nitrogen applications in Idaho crops. Bulletin 927.

Ruser, R., Flessa, H., Russow, R., Schmidt, G., Buegger, F., Munch, J., 2006. Emission of N2O, N2 and CO2 from soil fertilized with nitrate: effect of compaction, soil moisture and rewetting. Soil Biol. Biochem. 38, 263–274.

Santonja, G.G., Georgitzikis, K., Scalet, B.M., Montobbio, P., Roudier, S., Sancho, L.D., 2017. Best Available Techniques (Bat) Reference Document for the Intensive Rearing of Poultry or Pigs. EUR 28674 EN.

Schlesinger, W., Bernhardt, E., 2013. Biogeochemistry, third ed. Elsevier, New York.

Schlesinger, W.H., 2009. On the fate of anthropogenic nitrogen. Proc. Natl. Acad. Sci. U. S. A 106, 203–208.

Shahzad, A.N., Qureshi, M.K., Wakeel, A., Misselbrook, T., 2019. Crop production in Pakistan and low nitrogen use efficiencies. Nat. Sustain. 2, 1106–1114.

Sharma, L.K., Bali, S.K., 2018. A review of methods to improve nitrogen use efficiency in agriculture. Sustainability 10, 51.

Sharma, A., Behera, U., 2010. Green leaf manuring with prunings of Leucaena leucocephala for nitrogen economy and improved productivity of maize (*Zea mays*)–wheat (*Triticum aestivum*) cropping system. Nutr. Cycl. Agroecosyst. 86, 39–52.

Shaviv, A., 2001. Advances in Controlled-Release Fertilizers.

Shibata, H., Galloway, J.N., Leach, A.M., Cattaneo, L.R., Cattell Noll, L., Erisman, J.W., Gu, B., Liang, X., Hayashi, K., Ma, L., Dalgaard, T., Graversgaard, M., Chen, D., Nansai, K., Shindo, J., Matsubae, K., Oita, A., Su, M.-C., Mishima, S.-I., Bleeker, A., 2017. Nitrogen footprints: regional realities and options to reduce nitrogen loss to the environment. AMBIO 46, 129–142.

Signor, D., Cerri, C.E.P., 2013. Nitrous oxide emissions in agricultural soils: a review. Pesqui. Agropecuária Trop. 43, 322–338.

Signor, D., Cerri, C.E.P., Conant, R., 2013. N2O emissions due to nitrogen fertilizer applications in two regions of sugarcane cultivation in Brazil. Environ. Res. Lett. 8, 015013.

Smil, V., 2004. Enriching the Earth: Fritz Haber, Carl Bosch, and the Transformation of World Food Production. The MIT Press, Cambridge, MS, USA.

Sommer, S.G., Schjoerring, J.K., Denmead, O., 2004. Ammonia emission from mineral fertilizers and fertilized crops. Adv. Agron. 82, 82008-82004.

Stark, C.H., Richards, K.G., 2008. The continuing challenge of nitrogen loss to the environment: environmental consequences and mitigation strategies. Dyn. Soil Dyn. Plant 2, 41−55.

Subbarao, G., Ishikawa, T., Ito, O., Nakahara, K., Wang, H., Berry, W., 2006. A bioluminescence assay to detect nitrification inhibitors released from plant roots: a case study with Brachiaria humidicola. Plant Soil 288, 101−112.

Subbarao, G., Nakahara, K., Hurtado, M.d.P., Ono, H., Moreta, D., Salcedo, A.F., Yoshihashi, A., Ishikawa, T., Ishitani, M., Ohnishi-Kameyama, M., 2009. Evidence for biological nitrification inhibition in Brachiaria pastures. Proc. Natl. Acad. Sci. U. S. A 106, 17302−17307.

Subbarao, G., Sahrawat, K.L., Nakahara, K., Ishikawa, T., Kishii, M., Rao, I., Hash, C., George, T., Rao, P.S., Nardi, P., 2012. Biological nitrification inhibition—a novel strategy to regulate nitrification in agricultural systems. Adv. Agron. 114, 249−302.

Sutton, M., Dragosits, U., Theobald, M., McDonald, A., Nemitz, E., Blyth, J., Sneath, R., Williams, A., Hall, J., Bealey, W., 2004. The role of trees in landscape planning to reduce the impacts of atmospheric ammonia deposition. Landscape ecology of trees and forests. In: Proceedings of the Twelfth Annual IALE (UK) Conference, Cirencester, UK, 21−24 June 2004. International Association for Landscape Ecology (IALE (UK)), pp. 143−150.

Sutton, M.A., Bleeker, A., Howard, C., Erisman, J., Abrol, Y., Bekunda, M., Datta, A., Davidson, E., De Vries, W., Oenema, O., 2013. Our Nutrient World. The Challenge to Produce More Food and Energy With Less Pollution. Centre for Ecology and Hydrology.

Sutton, M., Raghuram, N., Kumar Adhya, T., Baron, J., Cox, C., de Vries, W., Hicks, K., Howard, C., Ju, X., Kanter, D., 2019. The Nitrogen Fix: from Nitrogen Cycle Pollution to Nitrogen Circular Economy, pp. 52−65. Frontiers 2018/2019: Emerging Issues of Environmental Concern.

Swensson, C., 2003. Relationship between content of crude protein in rations for dairy cows, N in urine and ammonia release. Livest. Prod. Sci. 84, 125−133.

Swierstra, D., Braam, R., Smits, M., 2001. Grooved floor system for cattle housing: ammonia emission reduction and good slip resistance. Appl. Eng. Agric. 17, 85−90.

Teixeira, R.B., Borges, M.C., Roque, C.G., Oliveira, M.P., 2016. Tillage systems and cover crops on soil physical properties after soybean cultivation. Rev. Bras. Eng. Agrícola Ambient. 20, 1057−1061.

Thomas, C.L., Acquah, G.E., Whitmore, A.P., McGrath, S.P., Haefele, S.M., 2019. The effect of different organic fertilizers on yield and soil and crop nutrient concentrations. Agronomy 9, 776.

Ti, C., Xia, L., Chang, S.X., Yan, X., 2019. Potential for mitigating global agricultural ammonia emission: a meta-analysis. Environ. Pollut. 245, 141−148.

Tiwari, J.K., Devi, S., Ali, N., Buckseth, T., Moudgil, V., Singh, R.K., Chakrabarti, S.K., Dua, V., Kumar, D., Kumar, M., 2017. Genomics Approaches for Improving Nitrogen Use Efficiency in Potato. The Potato Genome. Springer, pp. 171−193.

Tonitto, C., David, M., Drinkwater, L., 2006. Replacing bare fallows with cover crops in fertilizer-intensive cropping systems: a meta-analysis of crop yield and N dynamics. Agric. Ecosyst. Environ. 112, 58−72.

Tremblay, N., Wang, Z., Cerovic, Z.G., 2012. Sensing crop nitrogen status with fluorescence indicators. A review. Agron. Sustain. Dev. 32, 451−464.

Venterea, R.T., Maharjan, B., Dolan, M.S., 2011. Fertilizer source and tillage effects on yield-scaled nitrous oxide emissions in a corn cropping system. J. Environ. Qual. 40, 1521−1531.

Venterea, R.T., Halvorson, A.D., Kitchen, N., Liebig, M.A., Cavigelli, M.A., Grosso, S.J.D., Motavalli, P.P., Nelson, K.A., Spokas, K.A., Singh, B.P., Stewart, C.E., Ranaivoson, A., Strock, J., Collins, H., 2012. Challenges and opportunities for mitigating nitrous oxide emissions from fertilized cropping systems. Front. Ecol. Environ. 10, 562−570.

Vitousek, P.M., Aber, J.D., Howarth, R.W., Likens, G.E., Matson, P.A., Schindler, D.W., Schlesinger, W.H., Tilman, D.G., 1997. Human alteration of the global nitrogen cycle: sources and consequences. Ecol. Appl. 7, 737−750.

Vitousek, P.M., Menge, D.N.L., Reed, S.C., Cleveland, C.C., 2013. Biological nitrogen fixation: rates, patterns and ecological controls in terrestrial ecosystems. Philos. Trans. R. Soc. Lond. B Biol. Sci. 368, 20130119-20130119.

Vogeler, I., Hansen, E.M., Nielsen, S., Labouriau, R., Cichota, R., Olesen, J.E., Thomsen, I.K., 2020. Nitrate Leaching from Suction Cup Data: Influence of Method of Drainage Calculation and Concentration Interpolation. Wiley Online Library.

Wakeel, A., 2015. Balanced Use of Fertilizers in Pakistan: Status and Perspectives. Potash Institute (IPI) Switzerland, University of Agriculture Faisalabad, Pak.

Westhoek, H., Lesschen, J.P., Rood, T., Wagner, S., De Marco, A., Murphy-Bokern, D., Leip, A., van Grinsven, H., Sutton, M.A., Oenema, O., 2014. Food choices, health and environment: effects of cutting Europe's meat and dairy intake. Global Environ. Change 26, 196−205.

Wilkins, R.J., 2008. Eco-efficient approaches to land management: a case for increased integration of crop and animal production systems. Phil. Trans. Biol. Sci. 363, 517−525.

Yang, L., Bai, J., Liu, J., Zeng, N., Cao, W., 2018. Green manuring effect on changes of soil nitrogen fractions, maize growth, and nutrient uptake. Agronomy 8, 261.

Yao, Y., Gao, B., Zhang, M., Inyang, M., Zimmerman, A.R., 2012. Effect of biochar amendment on sorption and leaching of nitrate, ammonium, and phosphate in a sandy soil. Chemosphere 89, 1467−1471.

Yao, Y., Zhang, M., Tian, Y., Zhao, M., Zhang, B., Zhao, M., Zeng, K., Yin, B., 2018. Urea deep placement for minimizing $NH_3$ loss in an intensive rice cropping system. Field Crop. Res. 218, 254−266.

Zaman, M., Nguyen, M.L., Blennerhassett, J.D., Quin, B.F., 2008. Reducing NH3, N2O and NO3 - ,N losses from a pasture soil with urease or nitrification inhibitors and elemental S-amended nitrogenous fertilizers. Biol. Fertil. Soils 44, 693−705.

Zanatta, J.A., Bayer, C., Vieira, F.C., Gomes, J., Tomazi, M., 2010. Nitrous oxide and methane fluxes in South Brazilian Gleysol as affected by nitrogen fertilizers. Revis. Brasil. Ciênc. Solo 34, 1653−1665.

Zellweger, C., Steinbrecher, R., Laurent, O., Lee, H., Kim, S., Emmenegger, L., Steinbacher, M., Buchmann, B., 2019. Recent advances in measurement techniques for atmospheric carbon monoxide and nitrous oxide observations. Atmosp. Meas. Tech. 12, 5863−5878.

Zhao-Liang, Z., 2000. Loss of fertilizer N from plants-soil system and the strategies and techniques for its reduction [J]. Soil Environ. Sci. 1.

Zia, M.S., Aslam, M., Gill, M.A., 1992. Nitrogen management and fertilizer use efficiency for lowland rice in Pakistan. Soil Sci. Plant Nutr. 38, 323−330.

# Pathways to sustainable nitrogen use and management in Pakistan

**Muhammad Arif Watto[1], Tariq Aziz[1], Abdul Wakeel[2], Waqar Ahmad[3], Abdul Jalil Marwat[4], Munir Hussain Zia[5]**

[1]*University of Agriculture Faisalabad, Sub-Campus at Depalpur, Okara, Punjab, Pakistan;* [2]*Institute of Soil and Environmental Sciences, University of Agriculture Faisalabad, Punjab, Pakistan;* [3]*Asian Soil Partnership, and School of Agriculture and Food Sciences, The University of Queensland, Brisbane, QLD, Australia;* [4]*National Fertilizer Development Centre, Islamabad, Pakistan;* [5]*R&D Department, Fauji Fertilizer Company Limited, Rawalpindi, Punjab, Pakistan*

## 9.1 Nitrogen challenges and responses

Nitrogen (N) is a critical component of the global economy, food security, and environmental sustainability (Houlton et al., 2019). Conversely, the imbalanced and excessive use of N can negatively impact the global economy, lead to climate change, and degrade downstream ecosystems (details in Chapter 1). Given the importance of N, it is being highlighted that the success of many Sustainable Development Goals (SDGs) hinges on global N solutions. However, the SDG indicators reveal that although N is relevant to almost every SDG, but barely visible anywhere in the SDGs expect in the SDG 14.1 on life below water. It has been proposed to adopt N use efficiency (NUE) or N losses into the SDGs but this has yet to happen. Hence, it is imperative to work on solving N challenges to achieve lasting benefits for (i) global food security; (ii) soil, air, and water quality; (iii) climate change mitigation; (iv) biodiversity conservation; and (v) ecosystem restoration (Houlton et al., 2019). Currently, the global N fixation at a rate of 150 million tonnes N $y^{-1}$ has already crossed its limits for anthropogenic N fixations (Steffen et al., 2015) and thus causing environmental damages in the form of air and water pollution, effects on greenhouse balance, and effects on ecosystem and soil. Nitrogen pollution has gained attention by the global science community and environmental policy experts and by various UN bodies including the United Nations Environment Programme (UNEP), the Global Environment Facility (GEF), the Global Partnership on Nutrient Management (GPNM), and the United Nations Economic Commission for Europe (UNECE). On March 15, 2019 during its 4th assembly the UNEP adopted a resolution recognizing the damaging effects of reactive N (Nr) as a formal first step at a global scale. During the UNEP-4 assembly, it was urged that all possible options

Nitrogen Assessment. https://doi.org/10.1016/B978-0-12-824417-3.00006-X

should be considered to facilitate and strengthen coordination at national, regional, and global levels for sustainable Nr management.

Subsequently, the UNEP launched a global campaign on sustainable nitrogen use and management during the Colombo Declaration on October 24, 2019. Following the Colombo Declaration, the South Asia Cooperative Environment Programme (SACEP) and its member countries (entire South Asia) in collaboration with the Global Partnership on Nutrient Management, and the International Nitrogen Management System (INMS) committed to be the part of the global campaign to halve N waste by 2030. It was also emphasized that an international mechanism will be developed for devising an integrated N management policy with actionable policy goals for sustainable N use. In this regard, various UN bodies will help policy makers at national, regional, and global levels through appropriate training and capacity building in order to develop extensive understanding of the dynamics of N cycle and potential actions for mitigating N impacts.

Halving N waste by 2030 is a very ambitious target, which, in turn, may contribute saving about $100 billion worth of N resources each year and thus contributing to the post-COVID-19 economic recovery (Sutton et al., 2021). Many countries have already started taking initiatives to combat the adverse impacts of N waste on air and water quality, climate change, and on biodiversity at national levels. However, most countries have so far fragmented and incoherent policies related to Nr emissions and to deal with the associated impacts (Morseletto, 2019). New integrated N management policies are urgently needed to compliment the world's struggle to achieve environmental sustainability targets as set in the form of different SDGs.

## 9.2 Nitrogen use and management status in Pakistan

In Pakistan, nitrogenous fertilizers are the most widely used synthetic fertilizers which make up about 80% of the total nutrient consumption by the crop sector and account for about 3% of the global N fertilizer consumption (IFA STAT 2020). Since 1961 to 2017, the use of N fertilizers has reported to be increased from 62.1 Gg y$^{-1}$ to 3439 Gg y$^{-1}$ with an enormous increase from 4.9 kg ha$^{-1}$ to more than 150 kg ha$^{-1}$ during the same years. Based on historical data, recent estimates show that N use has been continuously increasing, whereas NUE has witnessed gradually decreasing from 58% in 1961–65 to mere 23% in 2009–13 in Pakistan (Raza et al., 2018). The decreasing NUE has led to a massive concurrent N surplus from 0.17 Tg N y$^{-1}$ to 3.6 Tg N y$^{-1}$ during the same period. It has been estimated that the overall N surplus has been reached 3581 Gg N y$^{-1}$ and so are the gaseous N emissions with an estimated value of 1224 Gg N y$^{-1}$ in 2013 (Raza et al., 2018). Surplus N is available either for accumulation in soils, leaching to groundwater aquifers or to be lost to the environment.

Nitrogen use in agriculture sector has failed to achieve its potential in terms of N efficiency and recovery in Pakistan (Ali et al., 2015). Excessive and imbalanced use of N in the agriculture sector has led to significantly decreased NUE over the past few decades. Currently, the estimated 23% NUE in Pakistan is far below than the global average of NUE in crop sector (Raza et al., 2018). Due to inadequate N management practices, Pakistan's total N emissions have substantially increased over the past 2 decades with $N_2O$ emissions from 98.73 Gg $y^{-1}$ to 144.79 Gg $y^{-1}$, $NH_3$ emissions from 867.88 Gg $y^{-1}$ to 1311.21 Gg $y^{-1}$, and NOx emissions from 740 Gg $y^{-1}$ to 1 1139 Gg $y^{-1}$, respectively, from 2000 to 2014 (EC-JRC, 2019). In Pakistan, primary sources of N emissions are the agriculture and the livestock sector. Apart from the direct emissions, a significant amount of indirect emissions occur due to inefficient management practices both in the agriculture and in the livestock sector (Ijaz and Goheer, 2020; Raza et al., 2018). Nonetheless, Pakistan's institutional and regulatory framework does not offer any incentives to address the damaging effects of the excessive and imbalanced N use in the recent policy developments. Given this scenario, the institutional and regulatory setup requires modernization of all practices and procedures involved in N production to consumption encompassing all relevant sectors such as agriculture, livestock, transport, and industry to reduce N waste and emissions and increase NUE. Reducing N use waste and increasing NUE is highly desirable for Pakistan as it will help decrease many adverse environmental effects and will reap benefits for food production and the overall economy. Although some realization has started toward N management, it is yet an underdeveloped theme in Pakistan. There exists a plethora of national and provincial policies, regulations, and institutions on agricultural development and environmental management, only a few of them tend to recognize and tackle this issue on a limited scale.

## 9.3 Current policies related to nitrogen use in Pakistan

The Green Revolution of 1960s led to a widespread application of N fertilizers for crop production in Pakistan which has substantially increased from 4.9 kg $ha^{-1}$ in 1961 to more than 150 kg $ha^{-1}$ in 2017 (Raza et al., 2018). It is being reported that the increasing trends in N applications will continue to grow to meet the food requirements of the rapidly growing population over the next 40 years in developing countries as well as in Pakistan. Currently, N application rates and NUE considerably vary among countries and regions. Nitrogen use in Pakistan is significantly higher but NUE is significantly lower than many other countries in different parts of the world (Raza et al., 2018; Shahzad et al., 2019) suggesting an enormous potential for increasing crop production by increasing NUE without necessarily increasing N input use.

Pakistan is yet struggling to meet the growing food demands of the rapidly increasing population. Currently, Pakistan is ranked at number 88 out of 107 countries at the Global Hunger Index (IFPRI, 2020) indicating that Pakistan has to do a lot more to ensure food security in the country. Pakistan's 18% population is undernourished, 45% facing severe stunting, 15% wasting, and 30% is underweight (National Food Security Policy 2018; NIPS and ICF, 2018). Pakistan Vision 2025 and the National Food Security Policy strategize to reduce the number of food-insecure people to half and move toward zero hunger by 2030 (MoPDSI, 2013), which, of course, cannot be achieved without significantly improving agricultural production. There are two options to increase crop production: (i) by expanding the cultivated area; and (ii) by improving yield per unit land area. However, further expansion in the cultivated area is not a feasible option as Pakistan has already been doubled its cultivated area over the past 60 years. Although crop yield has also been increased by almost fivefold over this period, it is yet not comparable with many developing countries. Like other countries, the increased crop production is mainly attributed to the use of synthetic fertilizers in Pakistan as shown by the positive correlation between the fertilizer use and the crop production. However, the increased fertilizer particular N fertilizer applications have led to a significant N wastage in the agriculture sector.

Currently, there is no direct N management policy available in Pakistan. However, several indirect N-related policies exist and some indirect measures are being undertaken such as

- Balance use of fertilizers is being encouraged by promoting the use of K and P fertilizers through subsidies on the use of balanced fertilizers in the Punjab under its Agriculture Policy 2018.
- Attempts are being made to develop and introduce efficient N fertilizers in collaboration with private sector and few efficient N fertilizers have been introduced at a limited scale.
- Split application of N fertilizers is widely adopted by the farmers for its better NUE.
- Burning of crop residues has been strictly banned especially rice stubbles. In this regard, Punjab Policy on Controlling Smog 2017 is a notable step in the Punjab province.
- Large-scale upscaling of high efficiency irrigation systems especially drip irrigation is being in progress which have indirect effects to reduce leaching losses of $NO_3^-$.

## 9.3.1 Rethinking the national fertilizer policy (1989 and 2001)

In Pakistan, N fertilizers were introduced in 1952 primarily through imports. Later, the government of Pakistan realized that its reserves of natural gas could confer a comparative advantage for substituting the imported N fertilizers with the locally manufactured N fertilizers. Starting from the late 1950s and the early 1960s, the government started pursuing an "import-substituting" domestic fertilizer industrial development policy program. As a result, about 92% of N fertilizer consumed in Pakistan was being produced locally during 2019–20 (NFDC, 2021).

In the coming years, the growing production and use of fertilizers in Pakistan led to the development of various regulations, policies, and institutions. The Provincial Essential Commodity Act (PECA), which was promulgated in 1971, made fertilizer production and marketing a federal subject. Just 2 years later, the Punjab Fertilizer

(Control) Order of 1973 rendered the provincial management of fertilizers subservient to PECA which helped further strengthen the federal regulators. Besides these initiatives, various key organizations were also established in order to promote fertilizer use such as Fertilizer Research and Development (R&D) under the Directorate of Soil Fertility in the research wing of the Agriculture Department of the government of Pakistan. In 1977, the National Fertilizer Development Centre (NFDC) was established to address various issues related to production, imports, pricing, subsidies, and regulations (Ali et al., 2015). After 12 years of the establishment of the NFDC, the first fertilizer policy was announced in 1989 which was updated in 2001. The Fertilizer Policy of 1989 was primarily about ensuring regular gas supply to the domestic fertilizer industry and permitting duty-free import of machinery to attract the local producers of fertilizer in the country. Nothing much new was introduced in the 2001 version of the Fertilizer Policy except for revising the existing fuel and gas prices and import duties on machinery for fertilizer industry. It states as cited by Ali et al. (2015):

> *This policy intends to provide investors of new fertilizer plants a gas price that enables them to compete in the domestic market with fertilizer exporters from the Middle East so that indigenous production is able to support the agriculture sector's requirement by fulfilling fertilizer demand.*

This policy exclusively focuses on the production side while ignoring the demand side, i.e., distribution and utilization. The most prominent gap is that this policy does not offer any incentives for improving fertilizer use efficiency particularly NUE. Pakistan's current fertilizer policy is not in line with the global initiatives to reduce N waste. Conversely, it is one of the examples of the existing actions that have negative impacts on efforts against N losses mitigations.

### 9.3.2 National Climate Change Policy 2012 and agricultural and food security policies

Pakistan approved its first National Climate Change Policy (NCCP) in 2012 with the aim to including climate change policy in the mainstream policy agenda for sustainable development. The United Nations Development Programme (UNDP) extended support through the One-UN Joint Programme on Environment for the formulation of the NCCP that was operationalized in 2013. The policy proposed 10 strategic objectives with more than 120 policy measures for agriculture, livestock, forestry, and energy sectors. Promoting conservation of natural resources and long-term sustainability was one of the key objectives. Although Pakistan's contribution to the global greenhouse gas (GHG) emissions is not significant, the implementation framework reflects its seriousness toward combating climate change. The NCCP provides a comprehensive framework for the development of action plans for national efforts for adaptation and mitigation measures against climate change.

Similarly, the NCCP 2012 recognized N emissions from the agriculture and the livestock sector and intends to exploring ways and methods to reduce N (particularly

N$_2$O) emissions from the agricultural soils such as by changing the traditionally used fertilizer mix and increasing fertilizer use efficiency. However, this policy needs to be further updated to address the newly emerging issues associated with imbalance and excessive N use across different sectors. For this purpose, the fertilizer policy can be linked with the climate change policy and the agricultural policies. Because the Ministry of Climate Change (responsible for the development and implementation of NCCP) and the Ministry of National Food Security and Research (MoNFSR) (responsible for the development and implementation of agricultural and fertilizer policies) are two distinct ministries which work with different mandates. The proposed draft of the National Agriculture and Food Security Policy by the MoNFSR does not include any measures toward institutional, interdepartmental, or interministry collaboration for sustainable N management. The National Food Security Policy 2018 only emphasizes on the need to make fertilizers available on time and at affordable prices.

Pakistan, a highly vulnerable country to the impacts of climate change, must rethink its future interventions which do not guarantee sustainable N management. An immediate new national agricultural policy and a new national fertilizer policy are needed fully appreciating the negative impacts of N losses and suggesting actionable both short-term and long-term measures to reduce N wastage and improve NUE. Furthermore, it has also become important as Pakistan is a part of the Colombo Declaration—a global campaign to halve nitrogen waste by 2030. There is an immediate need to improve fertilizer N productivity, crop yields, and reducing N losses to the environment. This requires the development of a policy framework to encourage sustainable N fertilizer use through better management practices. Nitrogen pollution is the entire planet's issue which needs to be dealt through multidimensional joined-up approaches and coordinated N management efforts.

## 9.4 Way forward for sustainable nitrogen management

As discussed in Section 7.1, a number of anthropogenic activities have disturbed the natural N cycle and N pollution has already crossed planetary boundaries. Taking action on N pollution can offer multiple cobenefits like climate change mitigation, improved air and water quality, saving coastal and marine environment, biodiversity conservation, etc. In Pakistan, a national roadmap for sustainable nitrogen management should require coordinated efforts at multiple levels such as at policy, research, and outreach levels engaging multiple stakeholders in nitrogen science, environmental policy, agriculture, industry, transport, civil society and development organizations, and environmental conservation organizations both national and international working in the country.

As major sector responsible for N pollution is agriculture (crop husbandry and livestock), hence, sustainable management of N would focus more on these two prominent N use sectors. It is recognized that N management in agriculture is intimately linked to the entire food system; hence, a complete assessment and management of the entire food system is needed if ambitious sustainability goals are to be achieved.

Specific measures are required to decrease pathway-specific N losses as the loss mechanisms differ between ammonia volatilization, leaching, surface run-off, $N_2O$, and NOx emissions. Possible trade-offs in the effects of N losses, abatement/mitigation measures may require priorities to be set, i.e., which adverse effects should be addressed first. Policy guidance is necessary to inform policy and value different options within the local context.

## 9.5  Pakistan's needs

Pakistan needs to have a toolbox based on the framework proposed by Oenema (2019) for developing an integrated sustainable N use and management approach for developing efficient and effective N pollution mitigation.

1. A comprehensive N system analysis based on concept of operations (CONOPS) which should include the identification and quantification of N components, processes, flows, actors, interactions, and interlinkages within and between different N systems.
2. Awareness among the general population, policy advisors, and legislators for explaining the meaning, purpose, targets, and actions of an integrated N management approach.
3. Quantification of the differences between N inputs and outputs of a system and the components of that system.
4. Environmental footprints analysis and cost−benefit analysis of N use.
5. Information on all available N sources.
6. Coordination (knowledge sharing/best practices) between research scientists, farmers, policy makers, and private sector within the country and outside the country.
7. Identification and application of best N use and management practices.

The following specific recommendations within the context of Pakistan are developed based on discussions with experts in N science and policy and literature review.

### 9.5.1  Policy interventions

- A National Fertilizer Standard Development Board (NFSDB) should be established under the Ministry of Food Security and Agriculture. NFSDB should develop fertilizer standardization protocols at the national level, whereas provinces should monitor different products according to the protocols of NFSDB and take actions in case of any violations.
- A national "Soil Profiling and Fertilizer Data Repository" should be established where research scientists and nutrients manufacturers/distributors could upload

the data related to N and other fertilizers. At present, there is no system about recording of crop wise N purchase by the farmers to help estimate exact consumption by a specific crop in a specific region. Quality datasets recording various agricultural inputs and outputs and other socioeconomic and environmental aspects will facilitate research work on climate change impact assessments and productivity projection studies. Digitalization of soil survey and analysis of the main agricultural lands and the already available yield prediction and fertilizer calculation software developed by University of Agriculture, Faisalabad, should be updated and linked with other research organizations and private fertilizer industry. Both public sector and private sector data should be made available for cross-validation.

- Current subsidy practices both indirect (through subsidized gas supply to the fertilizer industry) and direct subsidies (through stickers and coupon system fertilizer purchase) should be reconsidered. Indirect subsidies are more promising as it can cause a decrease in price of fertilizers, rather to give subsidies on coupon basis per bags. First, subsides on N fertilizers should gradually be phased out. However, if there is any subsidy to be given, it should be linked with mandatory actions in support of N management practices and to maximize NUE. Currently, any subsidies on fertilizers is given on 50 kg bags only (weight basis), ruling out all other fertilizers especially the enhanced efficiency fertilizers which have different packing sizes. If any subsidy is to be given, it should be given on per Kg of N not on the total weight of the fertilizer in a bag. In terms of indirect subsidy to the industry, the given subsidy can be adjusted against their general sales tax (GST) payments instead of allocating billions of rupees in the budget for giving subsidies.
- Burning of all crop residues and agricultural biomass similar to rice stubbles and wheat straw should also be banned throughout the country with strict implementation. Initiatives should be taken in the form of incentives and taking mechanized measures to incorporate stubbles and straw into soil.
- Biomass energy production should be promoted through special research funds to develop technologies for the integration of biogass plants with grid systems and improvement of biogass plants to secure high methane purity. Subsidies should be given to the farmers for promotion of the integrated farming system showcasing clean and green energy solutions.
- Capacity of the livestock farmers through technical advice by the livestock extension about livestock feed and manure management to reduce N wastage at livestock farms should be developed. Similarly, the capacity of crop farmers should be enhanced to take advantage of the scientific findings of relevant research organizations.
- Capacity of research organizations and researchers should be developed to make reliable predictions of climatic parameters and assessment of the nutrient flows and the corresponding likely impacts on various crops and to develop appropriate adaptation measures.

- Quality assurance institutions/organizations should be legally strengthened and scientifically equipped to effectively determine and monitor the quality of various liquid fertilizers, biofertilizers, SSP, DAP, NP, etc. Quality control laboratories should be given sufficient budget for analytical grade chemicals, trained work-force, and regular maintenance and upgradation of the laboratory equipment. Such laboratories should be established at least at tehsil level.
- "One Nutrition" policy should be followed in consultation with all relevant stakeholders interlinking soil−crop−human−animal health under the broader theme of "Agriculture linked human nutrition."
- At national level, the investment in agricultural research should be increased at least to 0.4% of agriculture GDP to make it at least at par with other countries in the region.
- Provision of subsidy should be considered on soil conditioners and technologies to regenerate and recover soil health.
- Traditionally, fertilizers are broadcasted manually in Pakistan which is not an efficient application method. Like many other technologies such as laser land levelers and high efficiency irrigation technologies, smart fertilizer applications technologies should be promoted and up-scaled.
- Integrated N planning at field, farm, sectoral and regional levels (including addressing the trend toward concentration of intensive livestock and crop farms, often near cities), fostering improved NUE, reduced wastage of $N_r$ resources, and a cleaner environment with less N pollution.
- Model integrated farms (livestock+crop) should be established and recommendations for sustainable N management should be devised considering the available resources.
- Different factors that hamper crop/animal growth need to be addressed by optimizing crop/animal production through NUE.
- The registration of the Fertilizer and Agriculture Firms/industry should be linked with hiring of Professional Agriculturists, which should be followed in true letter and spirit. In this regard, various skill development programs like diplomas in Agriculture, Livestock should be initiated with an objective of creating trained manpower in this sector.
- Research projects/funds should be allocated for assessment studies, development projects for new crop varieties with better nutrient use efficiency, development of smart fertilizers, biofertilizers, and nanofertilizers.
- A public−private partnership program should be initiated to develop enhanced efficiency fertilizers for smart agriculture in the country. In this regard, a matching grant system can be introduced where an equivalent share should come from the government side provided that the private, e.g., fertilizer companies come forward and make commitments for spending on R&D.
- The Ministry of Climate Change (MCC) should devise strategies for improvement in its recording methodology of the GHG inventories for which the following actions are suggested:
  - Efforts should be made to establish GHG inventory management systems
  - GHG inventories should be updated on regular basis

- Measures should be taken to develop Tier-II and Tier-III coefficients
- Steps should be taken to prepare remote sensing data—based GHG inventories
- Funds should be allocated for research on "Green/Blue Ammonia" to help develop fertilizer industry "low emission and efficient fertilizers."

## 9.5.2 Research and development initiatives

- Research and development is immediately needed to assess the amount of N accumulation and movement within the soil profile, rivers, and water resources in areas where N application rates are excessive and imbalanced. Zones should also be identified where ground water table is shallow and high precipitation is received alongside high application rates of N (including organic inputs) by farmers so that advisory can be issued accordingly to farmers in high risk zone.
- The available fertilizer recommendations throughout the country are too general and are very old; hence, site-specific and crop-specific N fertilizer use recommendations should be developed and the development and use of enhanced efficiency N fertilizers should be encouraged. Furthermore, customized crop-specific agricultural and fertilizer use machinery should be engineered and promoted.
- The production technologies of various important crops also need to be revisited to make these adaptable to the climate change scenarios.
- Research on local development of machinery for crop residue management is the need of the hour; hence, reverse engineering of imported implements and development of new implements as per local demands should be initiated.
- 4R Plus combines the 4Rs of nutrient stewardship—the right source of fertilizer, applied at the right rate and at the right time, and in the right place—Plus in-field and edge-of-field conservation practices that can increase productivity, bolster soil health, and improve water quality. 4R Plus the combination of 4R nutrient stewardship along with the conservation practices should be implemented in true spirit where a considerable amounts of N-sources (fertilizers) could be saved, and hence application would be cost-effective. Site-specific fertilizer recommendations should be issued at farmers' door-step alongside 4Rs advice. For this, the network of soil testing laboratories extending to tehsil level can greatly help.
- Studies on cropping patterns and the emerging issues associated with excessive N applications should be initiated.
- Develop and introduce better breeds of livestock with higher milk and meat productivity and which are less prone to heat stress.
- Research on efficient N utilization by livestock sector including feed management, silage, feed pelleting; decreasing raw protein should be initiated. Similarly, better management of manure and slurry should be devised and tested at commercial scale.

- Increase nutrient N (as well as others) recovery and recycling from waste and other underutilized resources should be investigated on priority.
- Besides local testing of N-efficient fertilizer products (including that of green/blue ammonia from renewable, carbon-neutral energy sources) developed globally, new low cost N inhibitor compounds (urease and nitrification inhibitors) should be identified through exploitation of the local flora which would perform better under high temperature and calcareous soils.
- Besides the above recommended measures, well researched natural plant products or by-products including neem (*Azadirachta indica*), Karanj (*Pongamia glabra*), vegetable tannins, waste products of tea, mint oil, Japanese mint, and Mustard (*Brassica juncea* L.) can be promoted as economical sources of nitrification inhibitors to increase the NUE. However, no extensive research has been conducted on using these natural substances in the field except for some of recent reports of neem oil coating of urea fertilizer in Pakistan. Local flora should preferably be screened out for identification of compounds, which depict NI properties. Commercialization of such products should be encouraged through industry—academia linkages—based research funds by different public sector funding agencies and industry, etc.
  - GHG emissions should be studied in various cropping systems particularly in rice, wheat, and maize being the major contributors.
    - a) Remote sensing- and GIS-based studies should be conducted for the identification of hotspots of N emissions. Government of Pakistan may also consider seeking technical support from the "Carbon Mapper Satellites" initiative of the US Government in this regard.
  - Impact of N pollution on lichen's growth in the northern areas and coral reefs in coastal area should be studied.
  - For estimation of $NO_3$ leaching, sampling from the high risk areas should be taken for onward analysis and mapping purpose.
  - Matching nitrogen inputs to crop needs (also termed "balanced fertilization") and to livestock needs offers opportunities to reduce all forms of N loss simultaneously, that can help to improve economic performance at the same time.

### 9.5.3 Outreach campaigns

- An outreach center for environmental extension and governance should be established for environmental awareness and education at provincial level. This center should organize media campaigns, short-courses such as MOOCs, trainings, and workshops to create awareness about various environmental issues such as N challenges as well as its sustainable management and build capacity of stakeholders including youth and women.
- Development of technical brochures related to NUE, climate change should be done in the local languages and placed at official website of agriculture departments, which could further be considered for promotion via social media and print and electronic media.

- Awareness campaigns about N challenge and the possible measures to improve NUE should be initiated through social media and the national mainstream electronic and print media.
- Role of essential crop nutrients and fertilizers should be made part of the primary education with emphasis about their judicious use to safeguard the environment. This would also help raise awareness among the future farmers about the role of fertilizers in food security.
- Fertilizer industry and livestock industry should be made part of any such media campaign so that program effectiveness could be enhanced to the next scale.
- Civil society should be taken on board for these awareness campaigns.

## 9.6 Conclusions

Nitrogen plays a critical role in the global economy, food security, and environmental sustainability. However, in recent years, experts in environmental assessments and environmental policy have raised concerns about the negative impacts of the excessive use of N on human health, ecosystem, and environmental sustainability. Given the recently realized N challenges, it becomes imperative to include N in the mainstream policy agenda for sustainable development. Different policies in different countries have started recognizing the nitrogen challenge and suggesting measures for sustainable N use across different sectors. Recently, the United Nations Environment Programme (UNEP) has launched a global campaign on sustainable nitrogen use and to halve N use by 2030. Pakistan is also a part of the Colombo Declaration and the global campaign fully appreciating that the success of SDGs hinge on global N solutions.

Pakistan is a leading nitrogenous fertilizer consuming countries, but its average NUE is far below than the global average. Consequently, a massive amount of N goes a surplus which is lost to the environment in one or the other way. Currently, there is no direct N management policy available in Pakistan and the indirect N-related policies do not adequately address the impacts of N losses. Hence, Pakistan needs to develop an integrated N use and management comprehensive approach based on N system analysis, N communication, N balance, integrated assessment modeling of N footprints, and cost–benefit analysis, N information, relevant stakeholder engagement, and best N management practices. Furthermore, measures at policy, institutional, research and development, and outreach level are required to decrease N losses and improve NUE. For this purpose priorities need to be set up, i.e., which adverse effects and which sector should be addressed at priority.

# References

Ali, M., Ahmed, F., Channa, H., Davies, S., 2015. The role of regulations in the fertilizer sector of Pakistan. In: Paper Presented at Conference, August 9-14, 2015, Milan, Italy 211559. International Association of Agricultural Economists.

EC-JRC, 2019. European commission, joint research Centre (EC-JRC)/Netherlands environmental assessment agency (PBL). In: Emissions Database for Global Atmospheric Research (EDGAR), Release EDGAR v5.0 (1970−2015) of November 2019.

Houlton, B.Z., Almaraz, M., Aneja, V., Austin, A.T., Bai, E., Cassman, K.G., et al., 2019. A world of co-benefits: solving the global nitrogen challenge. Earth's Fut. 7, 865−872.

ICF, 2018. Pakistan Demographic and Health Survey 2017−18. Islamabad, Pakistan, and Rockville, Maryland, USA: NIPS and ICF. https://www.nips.org.pk/abstract_files/PDHS %20-%202017-18%20Key%20indicator%20Report%20Aug%202018.pdf.

IFA Stat, 2020. https://www.ifastat.org/.

IFPRI, 2020. Global Hunger Index 2020: Pakistan. https://www.globalhungerindex.org/pdf/en/2020/Pakistan.pdf. (Accessed 20 April 2021).

Ijaz, M., Goheer, M.A., 2020. Emission profile of Pakistan's agriculture: past trends and future projections. Environ. Dev. Sustain. 23, 1668−1687.

MoPDSI, 2013. Pakistan Vision 2025: One Nation−One Vision. Government of Pakistan, Islamabad. Planning Commission, Ministry of Planning, Development and Special Initiatives (MoPDSI). https://www.pc.gov.pk/uploa ds/vision2025/Vision-2025-Executive-Summary.pdf.

Morseletto, P., 2019. Confronting the nitrogen challenge: options for governance and target setting. Global Environ. Change 54 (2019), 40−49.

National Food Security Policy, 2018. Government of Pakistan. Ministry of National Food Security and Research, Islamabad. http://www.mnfsr.gov.pk/frmDetails.aspx.

NFDC, 2021. National Fertilizer Development Centre, Islamabad. http://www.nfdc.gov.pk/Web-Page%20Updating/oftkpnt.htm. (Accessed 20 April 2021).

Oenema, O., 2019. Principles of integrated, sustainable nitrogen management. Draft section for a Guidance Document. In: Discussed at the Workshop on Integrated Sustainable Nitrogen Management 30 September-01 October, Brussels, Germany.

Raza, S., Zhou, J., Aziz, T., Afzal, M.R., Ahmed, M., Javaid, S., Chen, Z., 2018. Piling up reactive nitrogen and declining nitrogen use efficiency in Pakistan: a challenge not challenged (1961−2013). Environ. Res. Lett. 13 (3), 034012.

Shahzad, A.N., Qureshi, M.K., Wakeel, A., Misselbrook, T., 2019. Crop production in Pakistan and low nitrogen use efficiencies. Nat. Sustain. 2, 1106−1114.

Steffen, W., Richardson, K., Rockstrom, J., Cornell, S.E., Fetzer, I., Bennett, E.M., Biggs, R., Carpenter, S.R., de Vries, W., de Wit, C.A., Folke, C., Gerten, D., Heinke, J., Mace, G.M., Persson, L.M., Ramanathan, V., Reyers, B., Sorlin, S., 2015. Planetary boundaries: guiding human development on a changing planet. Science 347 (6223), 1−10.

Sutton, M.A., Howard, C.M., Kanter, D.R., Móring, A., Raghuram, N., Read, N., 2021. The nitrogen decade: mobilizing global action on nitrogen to 2030 and beyond. One Earth 4 (1), 10−14.

# Further reading

MoF, 2018. Pakistan Economic Survey (2017−18). Ministry of Finance (MoF), Islamabad. Gov Pakistan. http://www.finance.gov.pk/survey/chapters_18/Economic_Survey_2017_18.pdf.

# Index

Printed in the United States
by Baker & Taylor Publisher Services